Cyber Physical Systems

Architectures, Protocols, and Applications

WIRELESS NETWORKS AND MOBILE COMMUNICATIONS

Dr. Yan Zhang, Series Editor
Simula Research Laboratory, Norway
E-mail: yanzhang@ieee.org

Broadband Mobile Multimedia: Techniques and Applications
Yan Zhang, Shiwen Mao, Laurence T. Yang, and Thomas M. Chen
ISBN: 978-1-4200-5184-1

Cognitive Radio Networks: Architectures, Protocols, and Standards
Yan Zhang, Jun Zheng, and Hsiao-Hwa Chen
ISBN: 978-1-4200-7775-9

Cooperative Wireless Communications
Yan Zhang, Hsiao-Hwa Chen, and Mohsen Guizani
ISBN: 978-1-4200-6469-8

Cyber Physical Systems: Architectures, Protocols and Applications
Chi (Harold) Liu and Yan Zhang
ISBN: 978-1-4822-0897-9

Delay Tolerant Networks: Protocols and Applications
Athanasios V. Vasilakos, Yan Zhang, and Thrasyvoulos Spyropoulos
ISBN: 978-1-4200-1108-5

Distributed Antenna Systems: Open Architecture for Future Wireless Communications
Honglin Hu, Yan Zhang, and Jijun Luo
ISBN: 978-1-4200-4288-7

The Future of Wireless Networks: Architectures, Protocols, and Services
Mohesen Guizani, Hsiao-Hwa Chen, and Chonggang Wang
ISBN: 978-1-4822-2094-0

Game Theory for Wireless Communications and Networking
Yan Zhang
ISBN: 978-1-4398-0889-4

The Internet of Things: From RFID to the Next-Generation Pervasive Networked Systems
Lu Yan, Yan Zhang, Laurence T. Yang, and Huansheng Ning
ISBN: 978-1-4200-5281-7

Millimeter Wave Technology in Wireless PAN, LAN and MAN
Shao-Qiu Xiao, Ming-Tuo Zhou, and Yan Zhang
ISBN: 978-0-8493-8227-7

Mobile WiMAX: Toward Broadband Wireless Metropolitan Area Networks
Yan Zhang and Hsiao-Hwa Chen
ISBN: 978-0-8493-2624-0

Orthogonal Frequency Division Multiple Access Fundamentals and Applications
Tao Jiang, Lingyang Song, and Yan Zhang
ISBN: 978-1-4200-8824-3

Physical Layer Security in Wireless Communications
Xiangyun Zhou, Lingyang Song, and Yan Zhang
ISBN: 978-1-4665-6700-9

Resource, Mobility, and Security Management in Wireless Networks and Mobile Communications
Yan Zhang, Honglin Hu, and Masayuki Fujise
ISBN: 978-0-8493-8036-5

RFID and Sensor Networks: Architectures, Protocols, Security and Integrations
Yan Zhang, Laurence T. Yang, and Jimlng Chen
ISBN: 978-1-4200-7777-3

Security in RFID and Sensor Networks
Yan Zhang and Paris Kitsos
ISBN: 978-1-4200-6839-9

Security in Wireless Mesh Networks
Yan Zhang, Jun Zheng, and Honglin Hu
ISBN: 978-0-8493-8250-5

Unlicensed Mobile Access Technology: Protocols, Architectures, Security, Standards, and Applications
Yan Zhang, Laurence T. Yang, and Jianhua Ma
ISBN: 978-1-4200-5537-5

WiMAX Network Planning and Optimization
Yan Zhang
ISBN: 978-1-4200-6662-3

Wireless Ad Hoc Networking: Personal-Area, Local-Area, and the Sensory-Area Networks
Shih-Lin Wu, Yu-Chee Tseng, and Hsin-Chu
ISBN: 978-0-8493-9254-2

Wireless Mesh Networking: Architectures, Protocols, and Standards
Yan Zhang, Jijun Luo, and Honglin Hu
ISBN: 978-0-8493-7399-2

Wireless Quality-of-Service: Techniques, Standards, and Applications
Maode Ma, Mieso K. Denko, and Yan Zhang
ISBN: 978-1-4200-5130-8

Cyber Physical Systems

Architectures, Protocols, and Applications

Edited by
Chi (Harold) Liu and Yan Zhang

CRC Press is an imprint of the
Taylor & Francis Group, an **informa** business

CRC Press
Taylor & Francis Group
6000 Broken Sound Parkway NW, Suite 300
Boca Raton, FL 33487-2742

Printed on acid-free paper
Version Date: 20151014

International Standard Book Number-13: 978-1-4822-0897-9 (Hardback)

Visit the Taylor & Francis Web site at
http://www.taylorandfrancis.com

and the CRC Press Web site at
http://www.crcpress.com

Contents

List of Figures

List of Tables

List of Contributors

Peter Christen
Australian National University
Australia

Daqing Gu
France Telecom Orange Labs
China

Prem P. Jayaraman
RMIT University, Australia

Srimal Jayawardena
CSIRO, Australia

Chi Harold Liu
Beijing Institute of Technology
China

Charith Perara
Open University, U.K.

Zhengguo Sheng
University of Sussex, U.K.

Hao Wang
France Telecom Orange Labs
China

Arkady Zaslavsky
CSIRO, Australia

Yan Zhang
Simula Research Laboratory
Norway

Jianxin Zhao
Beijing Institute of Technology
China

Chapter 1

Background

Chi Harold Liu and Jianxin Zhao

Beijing Institute of Technology, China

Yan Zhang

Simula Research Laboratory and University of Oslo, Norway

In the past decades, with the fast improvements in science, technology, and engineering, people are continuing to redefine the capabilities of the surrounding physical world. The articles "Cyber Physical Systems" (CPS [1, 2, 3]) and "The Internet-of-Things" [4, 5, 6]) represent an evolution in computerized interconnectivity. They integrate the dynamics of the physical processes with those of the software and communication, providing abstractions and modeling, design, and analysis techniques for the integrated whole [7]. It has become the new frontier research focus for worldwide industry, academia, and government agencies [6]. The concept of CPS can be traced back to the pioneering work done by Kevin Ashton in 1999, and it is initially linked to the new idea of using radio frequency identification (RFID [8, 9]) in the supply chain industry sector [10]. Soon after, this term became popular and now is well known as a new communication and computing system, where the Internet is connected to the physical world via ubiquitous (wireless) sensor networks [11]. CPS's basic concept is that "...things having identities and virtual personalities operating in smart spaces using intelligent interfaces to connect and communicate within social, environmental, and user contexts..." [12].

These global *networked* things include RFID tags, ZigBee sensors [13], smartphones, etc., thus forming intimate coupling between the cyber and physical. Using technologies like wireless sensor networks (WSNs) [14], RFID [9],

ZigBee, WiFi and the next generation of mobile communication networks like 4G/LTE [15]. It is expected that various access networks, sensors/actuators, and almost everything on the planet can be seamlessly inter-connected together and be able to perceive and understand the complex surrounding environment, and hence cover a wide range of applications like healthcare [16, 17], home automation [18], environmental monitoring [19], intelligent transportation [20], electric power grid [21], utilities, e-learning [22], etc. [23]. Future applications include using such capabilities to enable highly intelligent and responsive actuation, for example through dynamic public transit scheduling or efficient and predictive utility (electricity, water) management. In recent years, with the evolving model of "participatory sensing" [24, 25], the "sensors" in a CPS sensory environment are not only traditional embedded systems such as a sensing node/unit, but also the affordable, wireless, and easily programmable mobile devices such as smartphones and tablets. These new types of smart devices are equipping with many types of sensors like accelerometer, GPS, gyroscope, magnetometer, luxmeter, microphone, proximity sensor. Today mobile phones have become a ubiquitous central computing and communication device in people's lives [25]. The mobile device market is growing at a frantic pace and it won't be long before it outnumbers the human population. It is predicted that mobile phones combined with tablets will exceed the human population by 2017 [26, 27].

The vision of CPS is to allow *"things"* to be connected anytime, anyplace, with anything and anyone, ideally using any path, any network, and any service [28]. In order to realize this vision, we need a common operating platform, namely the middleware, that is scalable and supports a high level of interoperability. This platform enables sensor data collection, processing, and analysis. Efficient and feature-rich CPS middeware platforms are key enablers of the CPS paradigm. We are currently observing an emerging trend in middelware solutions that enable CPS [29]. However, most of the solutions are designed and developed to be used in the cloud environments where abundant resources are available. It is believed that middleware solutions designed specifically for low powered resource constrained computation devices are critical in order to realize the vision of CPS.

From another perspective, the Internet of Things (IoT) is part of the future Internet and will comprise billions of intelligent Internet connected objects (ICO) which will have sensing, actuating, and data processing capabilities. With the development of IoT technologies in the past few years, a wide range of intelligent and tiny sensing devices has been massively deployed in a variety of vertical applications and several major standardization alliances have gradually formed based on the interests of technology selections and commercial markets. Generally, sensing devices have common features of constrained energy resources, limited processing capability, vulnerable radio conditions, real-time nature of applications and no direct human interaction, etc. By interconnecting these devices using low-cost wireless communication technologies, usually called wireless sensor networks (WSN), a new ecosystem with a large deployment of smart applications has been formed. Each ICO will have one or more embedded sensors that will capture potentially enormous amounts of data. Some ICOs are large, and some are small. Similarly, some ICOs are expensive while others are cheap. The number of sensors deployed around the world is increasing at a rapid pace. These sensors continuously generate enormous amounts of data. However, collecting data from all the available sensors does not create additional value unless they are capable of providing valuable insights that will ultimately help to address the challenges we face every day (e.g., environmental pollution and traffic congestion management). Furthermore, it is also not feasible due to its large scale, resource limitations, and cost factors. When a large number of sensors are available from which to choose, it becomes a challenge and a time-consuming task to select the appropriate sensors that will help the users solve their own problems.

The challenges include CPS specific challenges such as privacy, participatory sensing, data analytics, GIS-based visualization and cloud computing apart from the standard WSN challenges including architecture, energy efficiency, security, protocols, and quality of service. Toward this end, the aim of this book is to introduce CPS from the standpoint of architectures, protocols, and applications. First, from Chapter 2 to Chapter 6, we give a whole picture of the architectures of CPS, and elaborate on it from perspectives including mobile sensing devices/platforms, naming/addressing/profile services, device search and selection, device management and energy management. Then, we introduce enabling technologies of CPS, from the fundamental networking technologies to machine-to-machine (M2M) communications and mobile cloud computing (MCC) in Chapters 7, 8, and 9. Finally, in Chapters 10, 11 and 12, we present a few representative CPS applications including connected healthcare, gaming in the public transport crowd, and MCC-enabled emerging CPS applications. Finally, we conclude this book in Chapter 13.

To conclude, CPS is a part of future Internet and ubiquitous computing. It envisions interactions between *things* that consist of sensors and actuators. As the price of sensors diminishes rapidly, we can soon expect to see very large numbers of *things*. As a very active research field, there are a variety of questions that need to be answered in CPS, at different layers of the architecture and from different aspects of systems design.

I

CPS ARCHITECTURE

Chapter 2

Overall Architecture for CPS

Chi Harold Liu and Jianxin Zhao
Beijing Institute of Technology, China

In this chapter, we discuss the overall CPS architecture. We analyze the different CPS applications from the bottom up and extract the common parts that form the vertical structure.

Throughout the introduction, a careful reader may ask what do CPS applications have in common. Although there exist CPS segments, such as intelligent transportation, smart logistics, and smart grid, etc., we notice that they all have a similar vertical structure, which is shown in Figure 2.1. The only difference is their technical solutions due to the nature of applications or other constraints. Nevertheless, this figure shows the overall CPS architecture considered in this book, from the physical phenomenon all the way up to the data center, and it is also widely adopted in industry. A number of manufacturers or service providers are located on part of this value chain to provide differentiated technologies and services. Generally, we can divide the CPS architecture into four layers:

- Physical/MAC Layer

  Physical/MAC Layer (or access networks) lays a foundation for CPS architecture. CPS uses devices (such as sensors or actuators) to sense or control the physical world. They are interconnected either wirelessly or wired through a variety of radio access network technologies

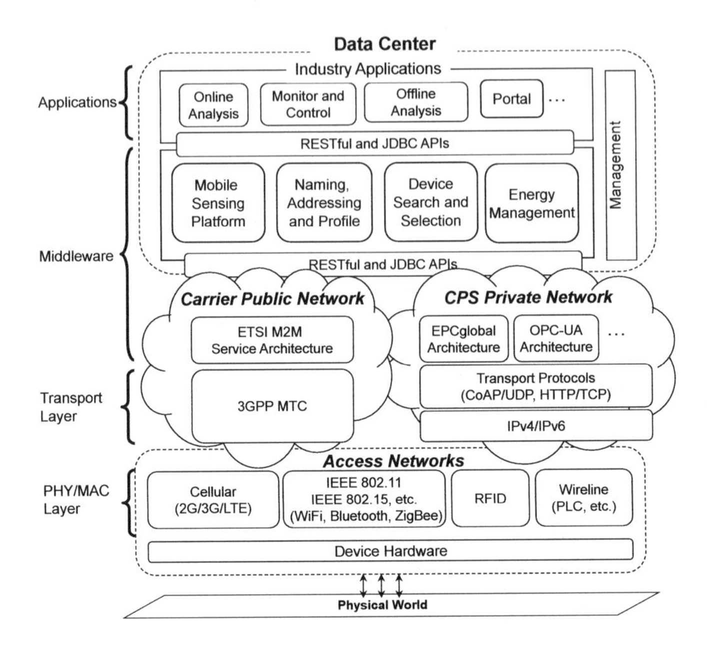

Figure 2.1: An overall architecture considered in this chapter.

for smart transmission. A few known examples are cellular networks (2G/3G/LTE), IEEE 802.11/802.15 series of standards for WiFi, ZigBee, Bluetooth, RFID readers and tags, and wire-line technologies like power-line communications (PLC), etc.

■ Transport Layer

Transport layer plays a key role in bridging sensing networks with carrier networks. It actually behaves as a gateway to translate the captured event from the sensor layer into a standard format, interconnecting the access networks and the backbone Internet. Data are then routed either through the carrier public network or CPS private network. For the former, a standard like 3GPP MTC is defined to upgrade the existing backbone cellular network to manage M2M devices; for the latter, most service layer architectures, e.g., EPCglobal RFID architecture and OPC-UA client-server model, leverage existing transport layer protocols such as CoAP [30] over UDP, and HTTP over TCP.

■ Middleware/Service Layer

CPS platform is the core element in CPS architecture. It actually provides operation functions such as terminal management, protocol conversion, route forwarding, and service authentication for CPS applications. Moreover, extra service functions can be added into the platform such as billing, account separation, and statistics for CPS applications. In this layer, ETSI M2M service architecture can inter-work with 3GPP MTC via interworking function (IWF) that enables seamless integration of M2M SC layer with cellular MTC devices. That is, M2M SCs can invoke and leverage cellular MTC functions to optimize and support better M2M services. Meanwhile, cellular MTC functions can be enhanced for M2M SCs. Companies like InterDigital and Juniper Networks have this kind of solution [31, 32]. The middleware layer consists of the mobile sensing platform, naming/addressing/profiling module, device search and selection module, and the energy management module.

- Application Layer

  Application supports various applications in multiple areas, such as smart city, smart home and smart vehicle, online analysis, environment monitoring and control, etc. From the point of view of a telecom operator, they all come from two major solutions: the telecom operator's solution is where telecom operators try to provide end-to-end CPS solutions, and "over the top" solutions, while manufacturers provide their own CPS services or applications over operators' networks which are only treated as a pipeline.

Chapter 3

Mobile Sensing Devices and Platforms for CPS

Charith Perera

Open University, U.K.

Prem P. Jayaraman, Srimal Jayawardena, Arkady Zaslavsky

Australian National University

Chi Harold Liu

Beijing Institute of Technology, China

Peter Christen

Australian National University

CONTENTS

The preceding chapter gives a whole picture of the CPS architecture. To discuss it in detail, in this chapter, we discuss mobile sensing platforms and their applications toward different but interrelated paradigms such as CPS, sensing as a service, and smart cities. We highlight and briefly discuss different types of mobile sensing platforms and functionalities they offer. Mobile sensing platforms are more often integrated with smartphones and tablet devices. The resource constrained nature of the mobile devices requires different types of designs and architectural implementations. We proposed a software-based mobile sensing platform called *Mobile Sensor Data Engine (MOSDEN)*. It is a plug-in-based scalable and extendible CPS middleware for mobile devices that provides an easy way to collect sensor data from both internal and external sensors. MOSDEN acts as an intermediary device that collects data from external sensors and uploads to the cloud in real-time or on demand. We evaluate MOSDEN in both stand-alone and collaborative environments. The proof of concept is developed on the Android platform.

3.1 Introduction

In this chapter, we present the design and implementation details of the proposed middleware solution named *Mobile Sensor Data Processing Engine* (MOSDEN). MOSDEN is designed to support sensing as a service model [33, 34] natively. Further, MOSDEN is a *true zero programming* middleware where users do not need to write program code or any other specifications using declarative languages. Our solution also supports both push and pull data streaming mechanisms as well as centralized and decentralized (e.g., peer-to-peer) data communication. We employ a plugin architecture so developers can develop plugins allowing MOSDEN to communicate with their sensor hardware. We utilize the ecosystem that is built around the Android platform to share and distribute plugins. We designed and developed MOSDEN in such a way that it is interoperable with other cloud-based middleware solutions such as GSN. Our pluggable architecture is scalable and promotes ease-of-use. We present results of evaluating the performance of MOSDEN using devices with different capabilities and resource constraints in order to validate MOSDEN's scalability and suitability toward CPS domain. Additionally, we also evaluated MOSDEN in a collaborative environment where multiple MOSDEN instances work together to accomplish distributed sensor data collection and processing [35].

Hereafter, the chapter is organized in the following way: Section 3.2 presents the background information on mobile sensing systems in CPS paradigm. Different strategies and patterns in which mobile sensing systems can be used are discussed and compared in Section 3.3. In Section 3.4, the proposed solution, Mobile Sensor Data Engine (MOSDEN), is presented. The implementation details of MOSDEN are presented in Section 3.5. Section 3.6 discusses the performance evaluations and lessons learnt. Open challenges and opportunities are discussed in Section 3.7. Concluding remarks are presented in Section 3.8.

3.2 Mobile Sensing in Internet of Things Paradigm

Both cyber physical systems (CPS) and Internet of Things (IoT) paradigms envision the integration of computation, networking, and physical processes [36]. Toward this, mobile sensing platforms play a significant role. These middleware platforms allow us to collect data from internet connected devices. Additionally they are capable of processing data locally or pushing it to the cloud for further processing. Further, these platforms sit in between cloud middleware platforms and low cost sensors and establish the communication between the two. Sensing platforms primarily concentrate on collecting data but sometimes support actuation tasks up to a certain level. In order to explain sensing as a service domain better, let us introduce Figure 3.1.

It illustrates how cloud-based CPS middleware platforms receive data from sensors through mobile sensing middleware platforms. All the data collected are available to users through sensing as a service model [33]. Typically mobile sensing middleware platforms are more powerful than sensors and much less powerful than cloud servers in terms of processing, communication, and storage capabilities. For example, sensors may be capable of performing communication using short range protocols such as Bluetooth. From sensors perspective, it is cheap to support Bluetooth where its radio consumes far less energy compared to other long-range communication protocols such as 3G

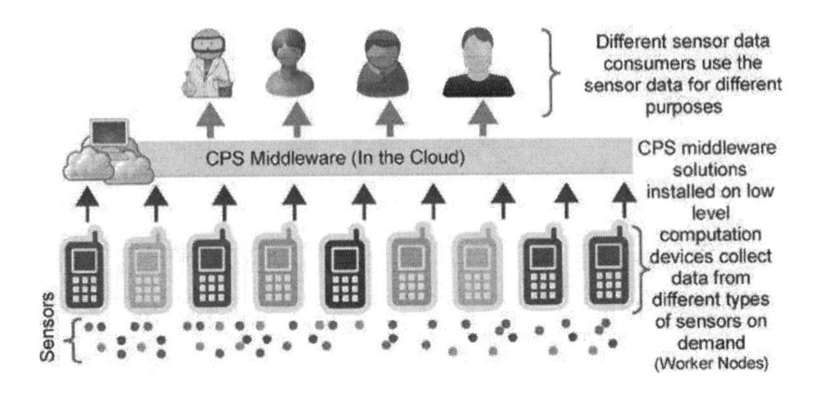

Figure 3.1: MOSDEN supports sensing as a service model. Sensors that do not have long-range network communication capabilities connect to MOSDEN instances. Then, MOSDEN processes the data and transmits it to the cloud selectively,

and WiFi. Short-range protocols also require less expensive radios (at the hardware level), which makes it more viable for sensors.

Computational devices with hardware configurations similar to modern smartphones can easily support long-range communication protocols such as 3G and WiFi. Therefore, this category of devices is ideal for sensing middleware platforms to be installed. As they have more computational resources they can perform intelligent processing, storage, filtering, and more. In this chapter, we define the mobility in two ways — mobility of sensors as well as mobility of sensing platforms. In Figure 3.2, we illustrate a number of devices that have similar capacity in terms of computational resources. MOSDEN is capable of supporting this category of devices. They are ranging from smartphones and tablets, low-cost computer devices such as Raspberry Pi, or wall-mounted tablets for smart home environments. If the mobile devices store data, MapReduce technique can be used to collect data required by the users [37].

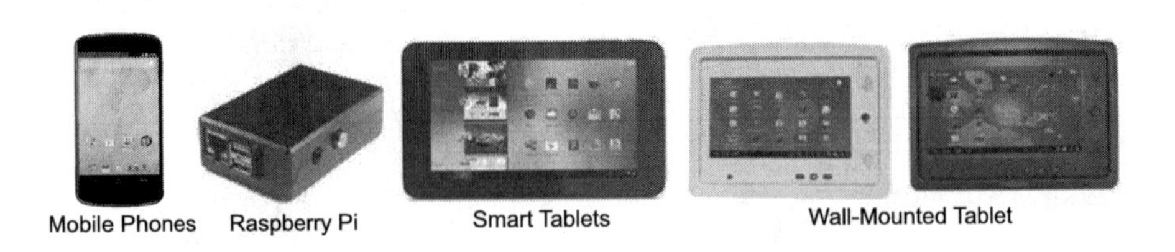

Figure 3.2: Different types of mobile sensing devices and platforms.

3.3 Strategies, Patterns, and Practice of Mobile Sensing

Choudhury et al. [38] has developed a custom mobile sensing hardware platform for activity recognition. Activities such as walking, running, taking stairs up/down, taking an elevator up/down, cooking, working on a computer, eating, watching TV, talking, cycling, using an elliptical trainer, and using a stair machine can be detected by using the device. Choudhury et al. have used sensors such as microphone, light, 3-axis digital accelerometer, barometer temperature, IR and visible+IR light, humidity/temperature, compass, 3D magnetometers, 3D gyroscope, and 3D compass to collect data to support their algorithms that detect the activities. Lee et al. [39] have developed a similar system. However, instead of processing the data in the mobile device, it sends data to the cloud by using a smartphone as an intermediate gateway device. Another similar approach has been presented by Laukkarinen et al. [40]. They have implemented a distributed middleware for 8-bit micro controller nodes where executing instructions (e.g., for data processing and event detection) are sent to each node using a *Process Description Language* (PDL). CONSORTS-S [41] has also used a similar approach. Instead of getting data from external sensors directly into mobile phones, CONSORTS-S uses a custom-made sensor board that connects to the mobile phone using a serial cable which allows the mobile phone to collect data from external sensors.

Most mobile sensing applications can be classified into *personal* and *community sensing* [25, 42]. *Personal sensing* applications focus on the individual. On the contrary, *community sensing* also termed *opportunistic/crowdsensing*[1] takes advantage of a population of individuals to measure large-scale phenomena that cannot be measured using a single individual. In most cases, the population of individuals participating in *crowdsensing* applications share a common goal. To date most efforts to develop *crowdsensing* applications have focused on building monolithic mobile applications that are built for specific requirements [43]. Further, the sensed data generated by the application are often available only within the closed population [44]. However, to realize the greater vision of a collaborative mobile *crowdsensing* application, we would need a common platform that facilitates easy development and deployment of collaborative crowd-sensed applications [45].

Grid-M [46] is a platform for lightweight grid computing. It is tailored for embedded and mobile computing devices. The middleware is built using Java 2 Micro Edition, and application programming interface (API) is provided to connect Java-developed applications in a grid computing environment. This work highlights the importance of providing an API-based communication channel which enables communication. As illustrated in Figure 3.1, mobile nodes work like the grid computing principle where they work together to collect sensor data as instructed by the cloud-based CPS middleware or by

[1] In this chapter, we use the terms *opportunistic sensing*, *crowdsensing*, and *participatory sensing* synonymously.

their own peers (e.g., other mobile sensing platform nodes). Zhang et al. [47] have developed a middleware on top of TinyOS (tinyos.net) for TelosB sensors. The data fusion components are designed as agents which migrate from one node to another. Such migration is an efficient technique in terms of resource utilization. Data fusion consumes the resources only when a given node is required to process data. Otherwise, the agent moves on to another node on demand. We simulate such behavior in MOSDEN where plugins are installed when needed and uninstalled when not needed. Another agent-based sensing platform has been proposed by Sun and Nakata [48]. Budde et al. [49] have proposed a framework that allows for smart objects discovery in the Internet of Things. The framework allows smart objects and services to be registered by providing metadata where it later allows searching and selection.

NORS [50] is an open-source platform that enables participatory sensing using mobile phones. It mainly focuses on collecting data instead of processing. The platform includes external sensors, mobile phones, and a cloud service for data storage. Sharing data among mobile phones is not supported. USense [51] is client-side middleware that opportunistically and passively (i.e., without human intervention) performs sensing tasks in crowd sensing fashion. It uses XML definitions to explain a *"moment"* where the middleware needs to start sensing and stop sensing. The *"moments"* are composed of a bunch of conditions such as location, time, and so on. Similarly, SENSE-SATION [52] also gathers and stores sensor information using mobile phones and makes them directly accessible over the Internet via RESTful web services.

MAGIC Broker 2 (MB2) [53] is a server-side platform for CPS paradigm where devices such as mobile phones and screens can be easily connected. MB2 allows objects interoperability by providing the API for queries.The platform provides basic abstractions, such as events, state, and content management services. Further, interoperability is supported via a middleware responsible for handling information coming from different devices where OSGimodularity plays a significant role in extensibility. SensorFly [54] is a controlled-mobile aerial sensor network platform for indoor, emergency response domain where it uses miniature helicopters to deploy sensors. When SensorFly is deployed, the nodes perform collaborative localization during flying. A very similar approach has been proposed in [55] where they use water flows to deploy sensors. TECO Envboard [49] is also a mobile sensor platform for urban sensing that is designed to be carried out by people or to be attached to other objects such as carrying bags, and bags in bicycles. The data collected is pushed to the cloud using a mobile phone. Wi2Me Traces explorer [56] is an extensible mobile sensing application for smartphones that allow any mobile user to gather not only access point locations but also their performance in terms of bandwidth, link quality, and successful connection rate. Such a platform is useful to diagnose issues in wireless sensor networks.

3.4 MOSDEN: Mobile Sensor Data Engine

In this section, we introduce the proposed solution in detail. First, we explain the problem that we addressed using the proposed solution. Then we illustrate the architecture of the solution followed by technical implementation details.

3.4.1 Problem Definition

We address several research problems in this work. Our focus areas are energy-efficient and effective data processing and network communication, cost-efficient infrastructure support for large-scale CPS deployment, and usability in connecting/configuring sensors. In the earlier section, we highlighted the importance of addressing the above-mentioned research challenges: (1) the importance of processing data locally before transmitting to the cloud, (2) the importance of utilizing devices with different computational capabilities and price tags, and (3) the importance of providing efficient and easy ways to connect sensors to low-level computational devices (devices belonging to category 3 and 4 in Figure 3.17).

There are several commercial solutions[2] that have been proposed in order to address some of the above-mentioned challenges. However, these solutions have several weaknesses. The following brief analysis helps to identify those weaknesses as well as to identify the ideal design requirements of a CPS middleware that needs to be installed on resource-constrained devices. Though some of the hardware components are open sourced, software systems remain closed source which makes it hard to extend and interoperate. Further, these solutions have their own hardware devices that perform tasks similar to MOSDEN. However, these devices are custom built. We believe utilizing commonly available devices, such as mobile phones, makes it easy to adopt due to the fact the most people are familiar with mobile phones and know how to operate them in comparison to custom-built proprietary devices. Another major drawback is inability for devices to interoperate with solutions provided by different vendors. For example, a sensor designed to be used by one solution cannot be connected to the software system of another solution. Hence, our proposed middleware aims to be vendor agnostic.

3.4.2 MOSDEN: Architectural Design

In this section, we explain the design decisions in detail. First, we present the reasons for introducing a plugin architecture. Second, we explain the complete MOSDEN architecture. Third, we explain how MOSDEN instances interact with their peers and GSN cloud instances. Finally, we briefly show that data can also be distributedly processed and MOSDEN instances.

[2]TWINE (supermechanical.com), Ninja Blocks (ninjablocks.com), and Smart Things (smartthings.com).

3.4.3 Plugin Architecture

In MOSDEN, we employed a plugin architecture [57] in order to support three main requirements: scalability, usability, and community-based development. A plugin is an independent software component that adds a specific feature to an existing software application. In MOSDEN, each plugin translates generic communication messages to sensor specific commands in order to enable communication between MOSDEN and a specific sensor. When an application supports plugins, it enables customization. Further, MOSDEN plugins can be installed and configured at run time.

Scalability: Due to the plugin approach, MOSDEN can virtually support any sensor in the world. Anyone can develop plugins that allow MOSDEN to communicate with given sensors. Further, plugins consume very small amounts of storage space (e.g., 25KB). Therefore, large numbers of plugins can be stored even in a resource limited mobile device. Furthermore, MOSDEN automatically removes unused plugins when the memory is running low. New plugins can be downloaded through application stores such as *Google Play* or directly as .apk files. Separation of plugins from the main MOSDEN application helps to reduce the size of the application and also promotes plug-n-play. Practically, at a given point in time, only a small number of plugins need to be installed in order to facilitate sensor communication though thousands of plugins would be available on applications stores. Finally, the plugin architecture allows us to improve MOSDEN in the future, specially in the directions of automated sensor discovery and plugin installation based on context information.

Usability: MOSDEN is convenient to use as it allows us to collect data from sensors without programming efforts. Users are only required to download the matching plugin from an application store. Due to standardized plugin structure, MOSDEN knows how to communicate with each plugin. For the user, all the technical complexities and details are hidden and happen autonomously behind the scene.

Community-based Development: Plugin architecture allows us to engage with developer communities and support a variety of different sensors through community-based development. Our software is expected to release as free and open source in the future. We provide the main MOSDEN application as well as the standard interfaces developers can use to start developing their own plugins to support different sensors. We provide a sample plugin source code where developers only need to add their code according to the guidelines provided. Plugin model support increasingly enables the number of sensors supported by MOSDEN. Plugins for MOSDEN can be downloaded via applications stores such as Google Play as illustrated in Figure 3.3.

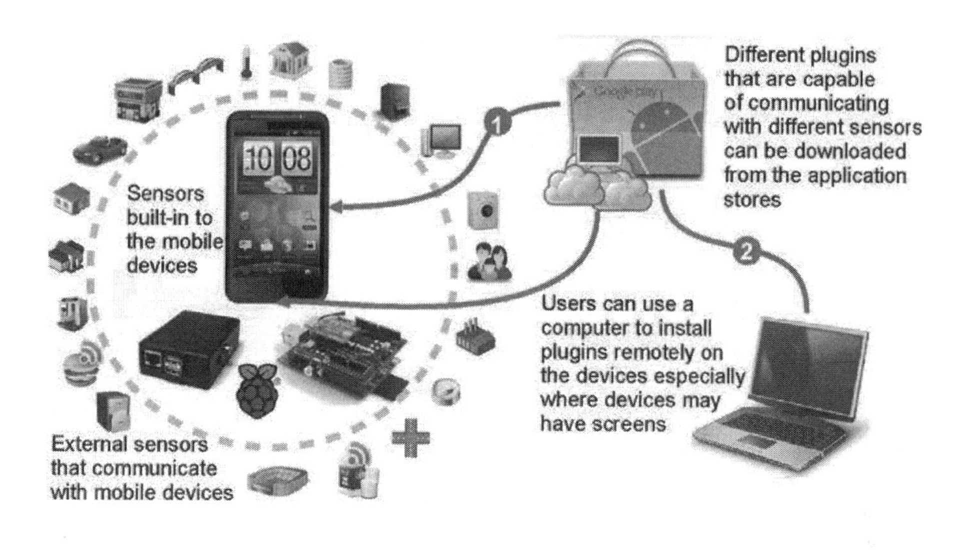

Figure 3.3: Plugin distribution and installation.

3.4.4 General Architecture

The architecture of MOSDEN is presented in Figure 3.4. MOSDEN architecture is based on the GSN architecture [58]. Additionally, we made several changes to the architecture in order to improve the efficiency as well as scalability. The major change is that we added a plugin manager and a plugin layer to support and manipulate plugins. GSN requires different wrappers to connect to different sensors. We eliminated this requirement and instead developed a single generic wrapper to handle the communication. In MOSDEN, wrappers do not directly communicate with sensors. Instead, the generic wrapper communicates with plugins and the plugin communicates with the sensors (i.e., wrapper → plugins (P_i)→ Sensor (S_i)). Due to the introduction of a generic wrapper, manual re-compilation of MOSDEN is not required when new sensors are added. Our newly added plugin manager component communicates with the cloud based GSN instances as well as MOSDEN peer instances and shares the information about the sensors connected to them. All the other architectural components behave as in the GSN middleware [58].

3.4.5 Interaction with the Cloud and Peers

MOSDEN is designed to be used as part of the sensing as a service model. On the other hand, due to that fact that our code is based on GSN middleware, MOSDEN is 100% compatible with GSN. This means communication between GSN instances and MOSDEN instances can be performed natively without any additional effort. Further, MOSDEN is a part of our overall vision of providing middleware support across different categories of devices as

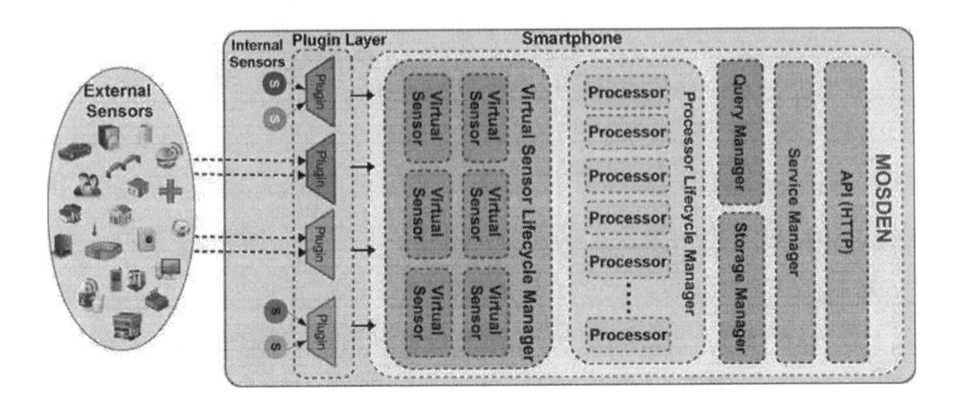

Figure 3.4: The architectural design of the MOSDEN. Legend: Sensor (S), Plugin (P), Wrapper (W), Virtual Sensor (VS). Plugins communicate with the sensors and retrieve data. Each plugin should be compatible with the sensor it wants to communicate with. Plugins compatible with different sensors can be downloaded from Google Play.

depicted in Figure 3.17. The typical interactions between GSN cloud instances and MOSDEN instances are illustrated in Figure 3.5. There are three main interactions that are frequently performed between MOSDEN instances and a GSN cloud instance. During our work, we also extended the cloud GSN architecture in order to support these interactions. When MOSDEN instance detects a new sensor connected to it through a plugin, it retrieves additional context information about the sensor (e.g., type of the sensor, unit measurements, manufacturer) from the sensor itself. Then, MOSDEN registers the newly detected sensor in the cloud GSN instance. Different MOSDEN instances register their own sensors independently in the cloud GSN instance. Cloud GSN combines all the information and models the data using the Semantic Sensor network ontology (www.w3.org/2005/Incubator/ssn/ssnx/ssn) [59, 60].

When the cloud GSN instance receives a request from a user, it queries the sensor description registry in order to find out the relevant sensors that match the user requirements. Then, it finds the MOSDEN instances that are capable of fulfilling the user request (i.e., whether the given MOSDEN is capable of collecting data from a sensor which is required by the user). Subsequently, the GSN instance sends the requests to MOSDEN instances. Then, each MOSDEN registers the request. Finally, MOSDEN starts streaming the requested data to the cloud GSN instance. The cloud GSN instance can make the requests in both pull and push mechanisms. In the pull method, GSN makes the request every time it wants data from MOSDEN. In the push method, cloud GSN sends the request and MOSDEN sends the data back until the request expires.

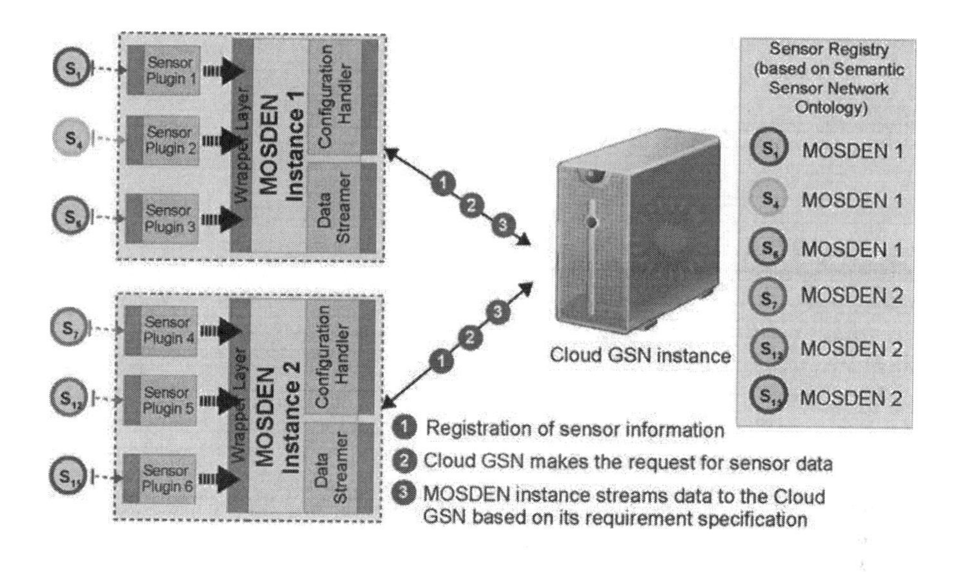

Figure 3.5: Interactions between MOSDEN and cloud GSN.

3.4.6 Distributed Processing

Though the topic of distributed data processing will not be discussed in detail in this chapter, it is important to note that processing data locally saves a substantial amount of network communication cost. Therefore, our proposed CPS middleware platform supports different resource-constrained mobile devices, where multiple MOSDEN instances can interact in peer-to-peer communication mode without having a central controller such as cloud MOSDEN. Everything discussed earlier can also be applied in distributed processing scenarios.

3.5 Implementation

In this section, we describe the implementation details of MOSDEN. First, we present an overview of the development platforms, tools, and technologies we used to develop the proposed solution. Further, we illustrate some user interfaces provided in MOSDEN. We also discuss how we implemented the plugin architecture and the steps and guidelines that need to be followed in order to develop new plugins that are compatible with MOSDEN.

Our middleware is written in Java and runs on Android-based devices. We used Java to develop our middleware in order to ensure compatibility with its cloud-based companion, GSN middleware [58]. Further, we selected Android platform due to its availability and the popularity.[3] Another important

[3]http://www.gartner.com/newsroom/id/2335616

factor is the portability of the Android platform. Android is not intended to be a platform only for mobile phones. The leading developer of the Android platform, Google Inc., intends to use it for many other smart devices such as automobiles, refrigerators, televisions, and so on. This vision supports our objectives discussed earlier in Section 3.4.1. Therefore, the objective of developing MOSDEN is not only to support mobile phone platforms but also to support devices such as *Raspberry Pi* (raspberrypi.org). Currently, Android for *Raspberry Pi* is under development. MOSDEN runs on Android 2.3 (and up), and it has 9935 (+ 768 logging lines in debugging version) lines of Java code. It consists of 115 classes distributed across 14 packages. MOSDEN is based on popular middleware called global sensor networks (GSN) [58]. MOSDEN source code will be available to download free in the future. As we mentioned earlier, our goal is not only to support mobile phones but also to support devices with similar resources limitations. These devices may or may not have screens. We decided to develop two different versions of our middleware based on the same underlying code-base where one version provides fully-fledged user interface to support direct user interaction with the MOSDEN as illustrated in Figure 3.6. The other version provides a simple user interface that only allows to start the middleware.[4] Figure 3.7 illustrates the user interface of the cloud GSN.

All the features available in GSN are also available in MOSDEN including data processing and REST-base peer-to-peer communication over HTTP. In comparison to GSN, we changed the wrapper structure and developed a generic wrapper. Further, we introduced the notion of plugins and added a plugin layer as well as a plugin manager. We also replaced the web-based user interface with a native Android application.

3.5.1 Plugin Development

This section explains how third party developers can develop plugins in such a way that their plugins are compatible with MOSDEN so MOSDEN can use them to communicate with external sensors. In plugin development, there are three main components that need to be considered: (1) Plugin interface written in Android Interface Definition Language (AIDL),[5] (2) plugin class written in Java, and (3) plugin definition in *AndroidManifest* file. Figure 3.8 shows the plugin interface written in AIDL. *IPlugin* is an interface defined in AIDL. Plugin developers should not make any changes in this file. Instead they can use this file to understand how MOSDEN plugin architecture works. *IPlugin* is similar to the Java interfaces. It defines all the methods that need to be implemented by all the plugins despite their functionalities. Related to MOSDEN,

[4] It is important to note that graphical user interface version requires Android 4.0 or higher as we have utilized the latest user interface components in order to provide a rich experience to the users. Limited user interface version is suitable for devices such as *Raspberry Pi* which reduces the additional overhead caused by the user interfaces.

[5] http://developer.android.com/guide/components/aidl.html

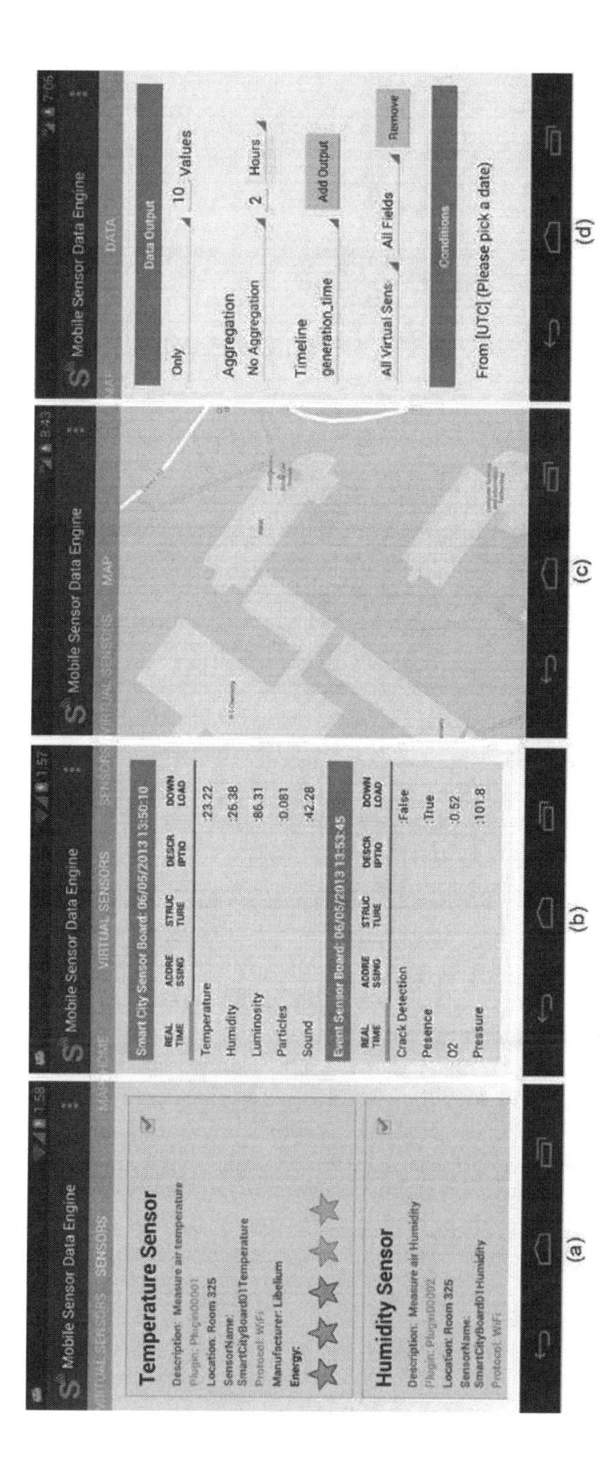

Figure 3.6: MOSDEN screenshots: (a) List of sensors connected to the MOSDEN; (b) list of virtual sensors currently running on the MOSDEN and their details; (c) map that shows sensor locations; and (d) interface for data fusing and filtering.

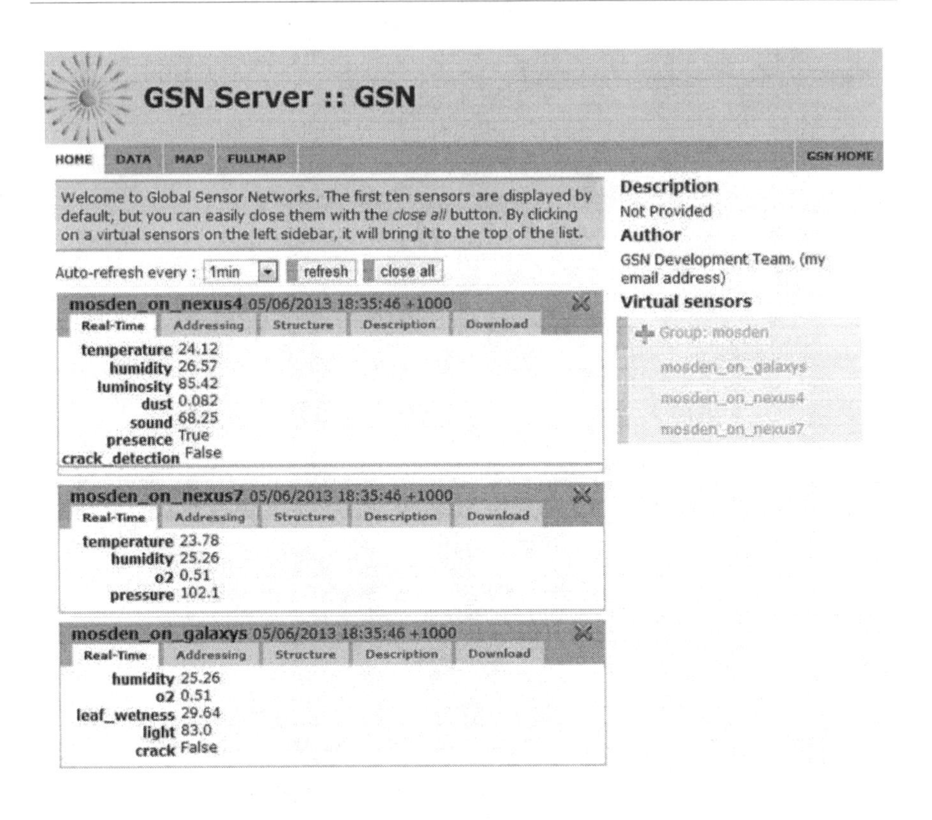

Figure 3.7: Screenshots of a cloud GSN instance showing three different MOSDEN instances registration.

we defined three methods to support the communication between main application and third party plugins.[6] Figure 3.9 presents the basic structure of a MOSDEN plugin. Each plugin is defined as an Android service. MOSDEN plugin developers need to implement these two methods: *getdataStructure()* and *getReadings()*. There is a third method, *void setConfiguration(in Map config)*, that developers can use to retrieve data from MOSDEN at runtime, especially information unknown to them at the development time (e.g., IP address, port number and other information related to configuration). This method accepts a Map[7] data structure as input and does not return any output.

In high-level, *getdataStructure()* returns a data type called *DataField4 Plugins[]*. This returning data structure describes what kind of data items MOSDEN should expect from the plugin, so MOSDEN can prepare its internal data structures as necessary. At the initialization phase, MOSDEN calls

[6]We expect to add more methods in order to support sophisticated functionalities and features in the future.

[7]A Java Data structure.

```
package au.csiro.mosden;

import au.csiro.mosden.beans.DataField4Plugins;
import au.csiro.mosden.beans.StreamElement4Plugins;

interface IPlugin {
  DataField4Plugins[] getDataStructure();
  StreamElement4Plugins[] getReadings();
  void setConfigurationInfo(in Map info);
}
```

Figure 3.8: IPlugin written in AIDL (Android Interface Definition Language) that governs the structure of the plugins. It defines the essential items in the plugin.

```
public class [Class] extends Service implements [Any Interface]{
   public int onStartCommand(Intent intent, int flags, int
   startId) {...}
   public void onDestroy() {...}

   public IBinder onBind(Intent intent) {...}

   private final IFunction.Stub mulBinder = new IPlugin.Stub(){
        public DataField4Plugins[] getDataStructure() throws
        RemoteException {...}

        public StreamElement4Plugins[] getReadings() throws
        RemoteException {…}

        public void setConfiguration(Map config) throws
        RemoteException {}
   }
}
```

Figure 3.9: MOSDEN plugin is an Android service.

the *getdataStructure()* method so MOSDEN knows what to expect before real data comes in. Once the initialization is done, MOSDEN calls *getReadings()* repeatedly depending on the frequency specified by the cloud GSN. The method *getReadings()* returns data raw (that comprise data items) that is organized as specified in the *DataField4Plugins[]*. The return data type is *StreamElement4Plugins[]*. Plugin developers are allowed to perform any operation within this method as long as it produces and returns the data types as specified by the guidelines.[8] Figure 3.10 shows how the plugins need to be defined in the AndroidManifest so MOSDEN application can automatically query and identify them. The Android plugin must have an intent filter

[8]We expect to release a developer guide that explains how third party plugins can be developed in the future.

```
<service
  android:name=[Plugin name]
  android:exported="true" >
  <intent-filter>
    <action android:name="au.csiro.mosden.intent.action.PICK_PLUGIN"/>
    <category android:name="au.csiro.mosden.intent.category.[PLUGIN_NAME]"/>
  </intent-filter>
</service>
```

Figure 3.10: Code snippet of the plugins *AndroidManifest* file.

```
<stream name="input1">
  <source alias="source1" sampling-rate="1" storage-size="1">
     <address wrapper="pluginwrapper">
        <predicate key="plugin">
             au.csiro.sensmalite.mainapp.intent.category.
             LIBELIUM_SMART_CITY_SENSOR
        </predicate>
        <predicate key="ip-address">130.56.73.110</predicate>
        <predicate key="port">20143</predicate>
     </address>
     <query>SELECT * FROM wrapper</query>
  </source>
  <query>SELECT * FROM source1</query>
</stream>
```

Figure 3.11: Code snippet of a virtual sensors definition.

and the action name must be *au.csiro.mosden.intent.action.PICK_PLUGIN*. Developers can provide any category name based on their preferences.

In order to support much user friendly and scalable plugin architecture, we extended the typical GSN Virtual Sensor Definition (VSD). The essential details that are required to connect a specific sensor to MOSDEN (e.g., IP address, port number) can be passed into the plugin via the VSD as illustrated in Figure 3.11. These details are important, especially in scenarios where multiple sensors need to use the same plugin (e.g., connecting 2 sensors that are similar).

3.6 Performance Evaluation and Lessons Learned

In this section, we present the details of the testbeds and evaluation methodology. We evaluate MOSDEN in both stand-alone and collaborative environments and results and lessons learned are discussed separately.

3.6.1 Experimental Testbed

We evaluated the proposed middleware solution, MOSDEN using several different parameters such as CPU consumption, scalability, memory requirements, latency and so on. For the evaluation, we used three devices with different resource limitations. From here onwards we refer them as D1, D2, D3, D4, D5, and D6. The technical specifications of the devices are as follows.

- **Device 1 (D1):** Google Nexus 4 mobile phone, Qualcomm Snapdragon S4 Pro CPU, 2 GB RAM, 16GB storage, Android 4.2.2 (Jelly Bean)
- **Device 2 (D2):** Google Nexus 7 tablet, NVIDIA Tegra 3 quad-core processor, 1 GB RAM, 16GB storage, Android 4.2.2 (Jelly Bean)
- **Device 3 (D3):** Samsung I9000 Galaxy S, 1 GHz Cortex-A8 CPU, 512 MB RAM, 16GB storage, Android 2.3.6 (Gingerbread)
- **Device 4 (D4):** Google Nexus 7 tablet, NVIDIA Tegra 3 quad-core processor, 1 GB RAM, 16GB storage, Android 4.2.2 (Jelly Bean)
- **Device 5 (D5):** Acer Iconia Tab A501, NVIDIA Tegra 2 T20 Dual-core 1 GHz Cortex-A9, 1 GB DDR2 RAM, Updated to Android 4.2.2 (Jelly Bean),
- **Device 6 (D6):** ASUS Ultrabook Intel(R) Core i5-2557M 1.70GHz CPU and 4GB RAM (Windows 7 operating system)

We used a computer D6 to host the cloud GSN during the evaluations. For our evaluations, we employed sensors built into the above devices (e.g., *Motion sensors:* accelerometer, gravity, gyroscope, liner acceleration, rotation vector; *Environmental sensors:* ambient temperature, light, pressure, relative humidity; *Position sensors:* magnetic fields, orientation, proximity.). Further, we used sensors manufactured by Libelium [61] as external sensors with different combination of hardware sensors plugged into them such as temperature sensor, humidity sensor, light-dependent resistor (LDR) sensor, air pressure sensor, leaf wetness sensor, noise sensor, dust sensor, force and pressure sensor, flex-bend sensor, flexible stretch sensor, hall-effect sensor, different gas sensors (e.g., O_2, CO_2) and so on. Resource constrained computational devices we used in this work as well as some of the sensors used in this experiments are shown in Figure 3.12.

3.6.2 Stand-Alone Experimentation

This section explains the evaluation methodology, experimental conditions and objectives of the Figures 3.13(a) to 3.13(g). All the evaluations are done using three different resource constrained mobile devices as explained in the section above. In all the evaluations, CPU usage (consumption) is measured in units

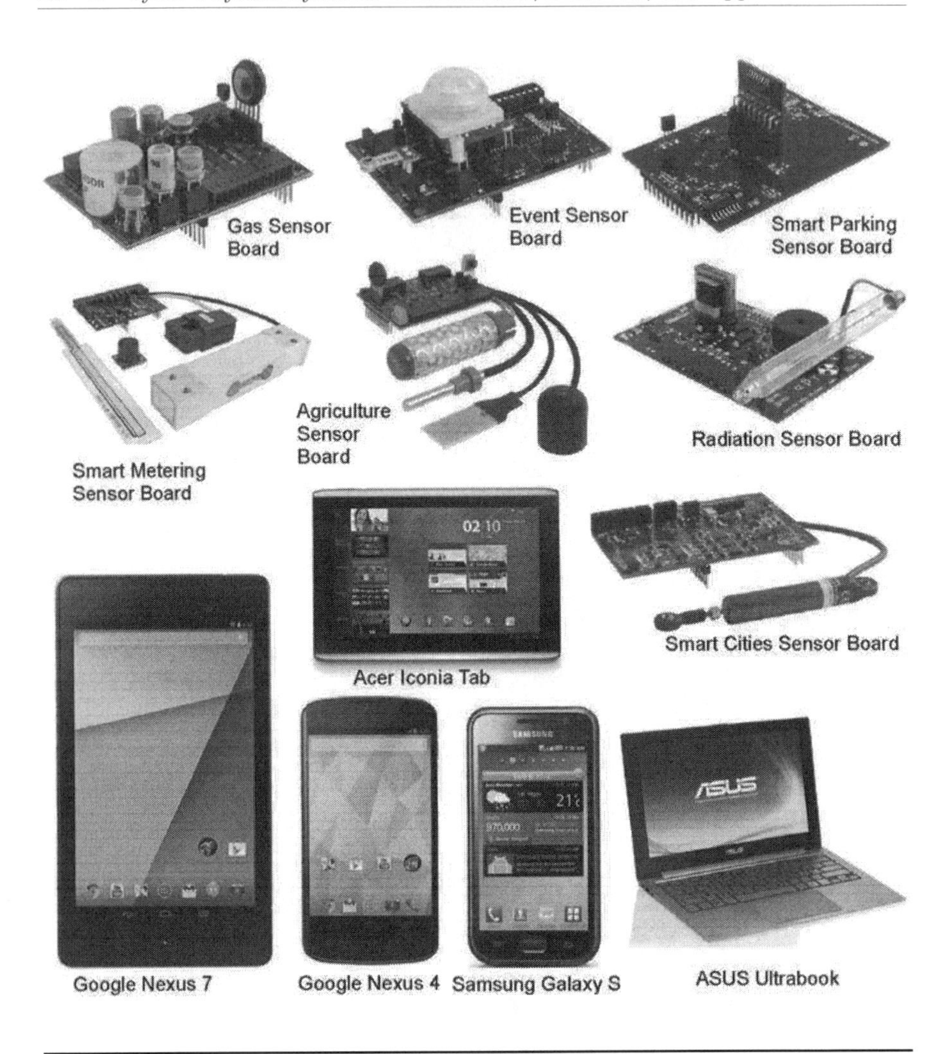

Figure 3.12: Some of the hardware devices used in the experimentation.

of *jiffies*.[9] At this point, MOSDEN supports only Wi-Fi communications.[10] We keep the sampling rate as 1 second during the course of evaluations.

In Figure 3.13(a), we examine how CPU usage changes when the number of sensors involved increases. Figure 3.13(b) shows how memory consumption changes when the number of sensors involved increases. Figure 3.13(c)

[9]In computing, a jiffy is the duration of one tick of the system timer interrupt. It is not an absolute time interval unit, since its duration depends on the clock interrupt frequency of the particular hardware platform.

[10]We expect to support ZigBee and Bluetooth in the future. However, such improvements will not change the overall architecture.

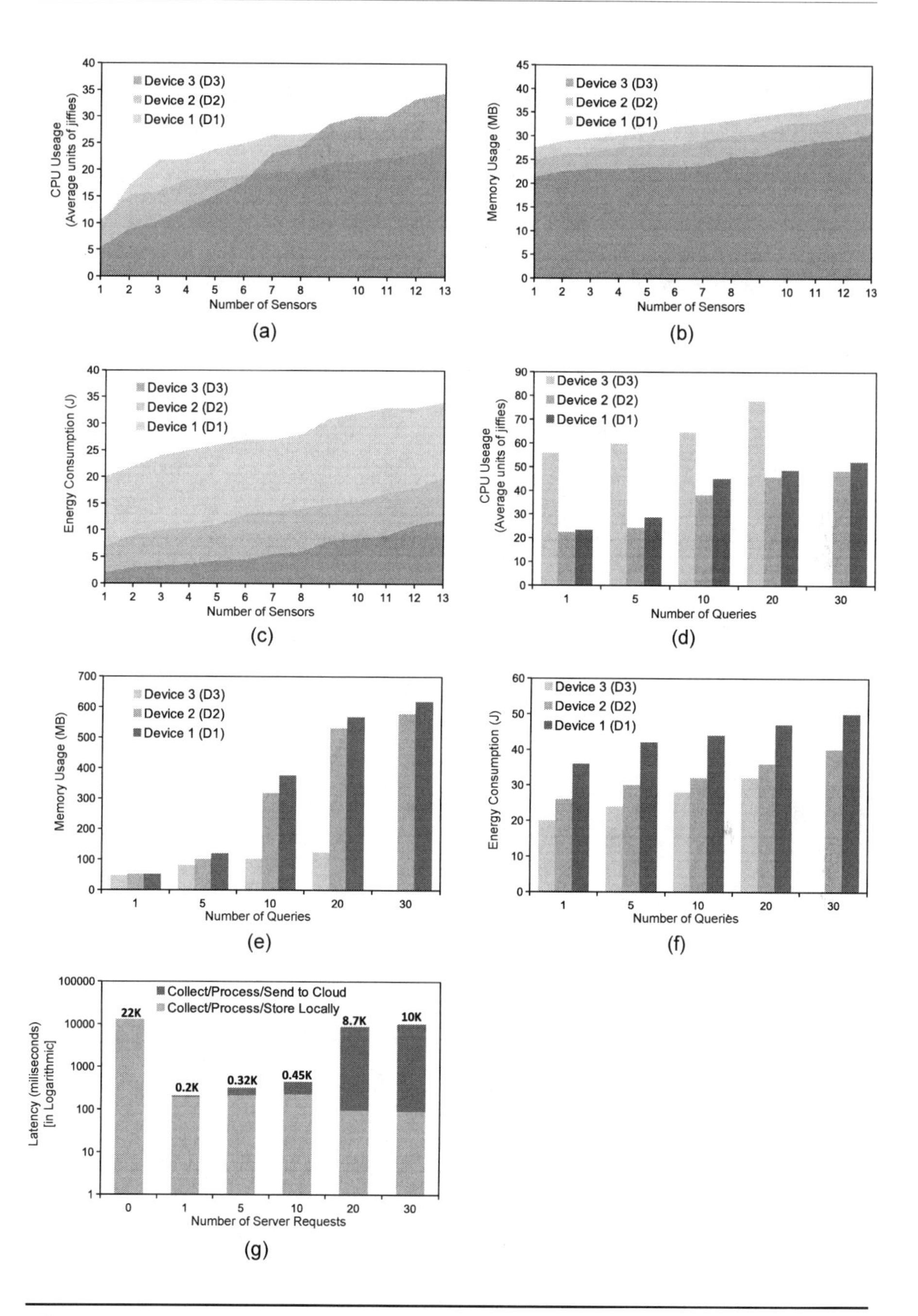

Figure 3.13: Experimentation and evaluation results. (Note: MOSDEN application and plugins use CPU, memory, and energy independently where plugins are treated as individual services by the Android platform. All the calculations are accumulated values of MOSDEN application and plugin services.)

measures how energy consumption changes when the number of sensors involved increases. In Figures 3.13(a), 3.13(b), and 3.13(c), MOSDEN only uses inbuilt sensors to collect data and store them in the local storage space. No network communication is performed. According to Figure 3.13(a), it is evident that CPU usage increases when the number of sensors increases. It is important to highlight that D3 consumes more CPU time compared to other devices when it needs to handle 10+ sensors. One reason for this is the lack of main memory (RAM) which puts additional overheads on the CPU.

Similar pattern is revealed in Figure 3.13(b) as well as in terms of memory usage. Devices that have larger memory capacity can afford to allocate more memory to MOSDEN, which increases the overall performance of MODSEN. Further, comparatively resource rich devices consume more energy due to usage of powerful CPUs and sensing hardware. This is observed in Figure 3.13(c) where difference in energy consumption for D1 and D3 is much higher compared to difference in memory usage. When not performing any network communication tasks, MOSDEN takes only 38MB (D1) / 30MB (D3) to collect, process and store data from 13 different sensors.[11] MOSDEN consumes around 35J (D1) / 10J (D3) to process, and store data from 13 sensors.

In Figure 3.13(d), we evaluate how CPU usage changes when the number of queries processed by the MOSDEN changes (step 2 and 3 in Figure 3.5). As we mentioned earlier related to Figure 3.13(a), Figure 3.13(d) also reveals that D3 uses significantly more CPU compared to other devices due to the overhead created by lack of memory. Comparatively, D1 and D2 use less CPU and as observed from the results, the CPU consumption is gradually increasing but not significant when MOSDEN processes more than 10 queries. One reason for this is that Android OS restricts MOSDEN from consuming too much CPU resource after a certain level as it needs to facilitate other essential Android applications and services.

Figure 3.13(e) shows how memory consumption changes when the number of queries changes. Additionally, Figure 3.13(f) shows how energy consumption changes when the number of queries changes. In Figures 3.13(d), 3.13(e), and 3.13(f), MOSDEN uses inbuilt sensors to collect data and send them to the cloud GSN over a WiFi network. Figure 3.13(e) clearly shows that D3 suffers from lack of memory as it is not allocated more than 150MB of memory. In contrast, both D1 and D2 have abundant memory available to be utilized so memory usage increases up to 620MB (D1) / 580MB (D2). The energy consumption graph with and without network communication looks similar in pattern. However, energy consumption has significantly increased across all three devices (50J (D1) / 40J (D2) when processing 30 queries).

In Figure 3.13(g), we examine the time MOSDEN takes (i.e., latency) to process and transmit the data. We measure the time taken for the following two operations. (1) We start measuring the time taken by the plugin to retrieve data from a sensor, pass it to a wrapper, and subsequently store it

[11] All the devices do not have all 13 sensors though the Android platform supports them.

in a local database. (2) We measure time taken for MOSDEN to respond to an incoming query request from the cloud GSN. When there is no network communication, MOSDEN takes 22 seconds to collect data from a sensor plugin, process, and store it locally. However, when the cloud GSN starts sending queries Android allocates more CPU and memory to MOSDEN. Hence, the data collection/processing and query processing operations are performed much faster which helps to reduce the overall latency from 22 seconds to 0.2 seconds. As the number of query requests increases, from the results, we observe that, latency[12] also increases. When MOSDEN processes 30 queries, latency increases to 10 seconds. However, a significant portion of the total processing time is taken to fuse the data and send them to the cloud.[13]

Lessons Learned: Our experimental evaluations validate the energy and performance efficiency of the proposed plugin-based MOSDEN platform. The middleware functioned without any issues during our experiments. Additionally, the plugin-based architecture increases the usability of MOSDEN by allowing users to download and install plugins from Google market place with zero effort in programming and no modifications to MOSDEN. Further, modern mobile devices can process significant numbers of requests with the limited resources they have. It is evident that the memory is more important than CPU in a situation where data needs to be processed under small sampling rates. In our previous work [62], we learned that reduced sampling rate can save energy and resource consumption significantly. In such scenarios, MOSDEN will be able to process many more queries efficiently than it did in the evaluations. We look forward to performing more experiments to examine the impact of sampling rate on MOSDEN's performance.

Potential Applications: The MOSDEN platform can greatly foster the development of new and innovative mobile data services that depend on CPS devices as the source of data. One such example is a crowd-sourcing application where sensor data (e.g., noise level in outdoor environments) can be collected from users' mobile devices running MOSDEN. The collected data can be used by applications in the cloud in their decision making process (e.g., determine the noise pollution level at an intersection in the city by fusing data from multiple MOSDEN instances). Another example is to determine real-time traffic conditions using data acquired from MOSDEN running on user mobile devices.

3.6.3 Collaborative Sensing Experimentation

In the previous section, we evaluated MOSDEN in stand-alone fashion. In this section, we evaluate MOSDEN in collaborative environment where multiple MOSDEN instances are configured to work together. For the collaborative evaluation, we used four mobile devices and a laptop. The technical specifi-

[12]Time it takes to fulfill all the requests made by the cloud GSN.
[13]Time that the data takes to travel over the network is not counted.

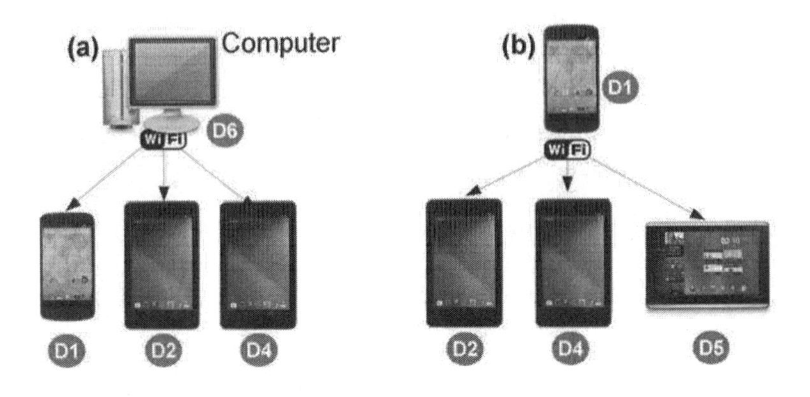

Figure 3.14: Experimental testbed has been configured in two different ways: (a) Setup 1: three mobile devices are connected to a laptop, and (b) Setup 2: three mobile devices are connected to another mobile device.

cations of the devices are listed in Section 3.6.1. In the following experiments we use D1, D2, D4, D5, and D6. For experimentation, we devised two setups as illustrated in Figure 3.14 and evaluated the proposed framework in each setup independently. The mobile devices are configured to run our proposed framework, MOSDEN, and the laptop computer is configured to run the GSN engine [63].

The overall objective of the experimental evaluations we conducted is to examine the performance of MOSDEN platform in collaborative environments. Two different collaborative setups are illustrated in Figure 3.14. In this section, we explain the objectives behind each experiment we conducted in detail. The next section discusses the results and lessons learned in detail. Number of sensors used for sensing has been kept fixed throughout the experiments.[14] In all the evaluations, CPU usage (consumption) is measured in units of jiffies.[15] Sampling rate for all evaluations is one second.

A query in the form of a *request* is sent from the server to MOSDEN client instances. Depending on the number of sensors queried on MOSDEN instances, the number of requests increases. We use the term *MOSDEN client* to refer to client devices where MOSDEN acts as a client such as D1, D2, and D4 in setup 1 in Figure 3.14(a) and D2, D4, and D5 in setup 2 in Figure 3.14(b)). We use the term *MOSDEN server* to refer to the server device where MOSDEN acts as a server such as D1 in setup 2 in Figure 3.14(b)).

We configured the experimental testbed as illustrated in Figure 3.14(a) - setup 1. In Figures 3.15, 3.15(b), and 3.15(c), we compare the performance

[14]All the sensors available on the given device has been used (e.g., in D1: accelerometer, microphone, light, orientation, proximity, gyroscope, magnetic, pressure).

[15]In computing, a jiffy is the duration of one tick of the system timer interrupt. It is not an absolute time interval unit, since its duration depends on the clock interrupt frequency of the particular hardware platform.

of *restful streaming* and *push-based streaming* methods in term of CPU usage and memory usage by both client and server devices which run MOSDEN and GSN. Restful streaming is designed to have a persistent connection between the client and the server. On the other hand, the push-based approach makes a new connection every time to transmit data. Both these techniques can be used to perform communication between two (or more) distributed GSN or MOSDEN instances (i.e., GSN $\leftrightarrow$ GSN, MOSDEN $\leftrightarrow$ MOSDEN, GSN $\leftrightarrow$ MOSDEN). The two approaches have their own strengths and weaknesses. The former is good for clients running MOSDEN that have a reliable data connection. The latter is useful for clients that need to work in offline modes. The MOSDEN platform supports both the operations and the application developer has the choice to choose the best approach suited to application requirements.

Figure 3.15(a) illustrates the difference between CPU usage in MOSDEN when number of requests increases. Figure 3.15(b) illustrates the variation of memory consumption of MOSDEN when number of requests increases. According to Figure 3.15(a), it is evident that restful streaming is slightly better than push-based streaming from a CPU consumption perspective. This slight difference can be due to above explained reasons. I contrast, restful streaming consumes more memory than push-based streaming as depicted in Figure 3.15(b). One reason could be the overheads to maintain a persistent network connection.

Figure 3.15(c) illustrates how memory consumption of GSN changes in the server when the number of queries it handles increases. It can also be noted that the memory consumption of GSN engine running on the server as depicted in Figure 3.15(c) also increases with load but not as significantly as the mobile device. This observation is straightforward and attributed to the difference in computing capacity of the two nodes (mobile device and laptop).

Based on the experience in MOSDEN client-side, it is fair to predict that, we will be able to see a difference if we increase the number of requests to be processed toward tens of thousands. In Figure 3.15(d), we examine how storage requirements vary when the number of sensors handled by the MOSDEN client increases. For this experiment, we used Setup 1 in Figure 3.14. All the sensors onboard the client mobile device (i.e., accelerometer, microphone, light, orientation, proximity, gyroscope, magnetic, pressure) are used as sensor sources. Sampling rates for sensors are configured as one second. The D1 (Setup 1) has been configured to receive data requests from the server at one-second intervals. The experiment was conducted for three hours. The exact storage requirements depend on multiple factors such as number of active sensors sending data, number of data items generated by the sensor,[16] sampling rate, and history size [58]. We used external sensors to increase the number

[16]E.g., accelerometer generates 3 data items i.e., x, y, and z while temperature sensor generates one data item.

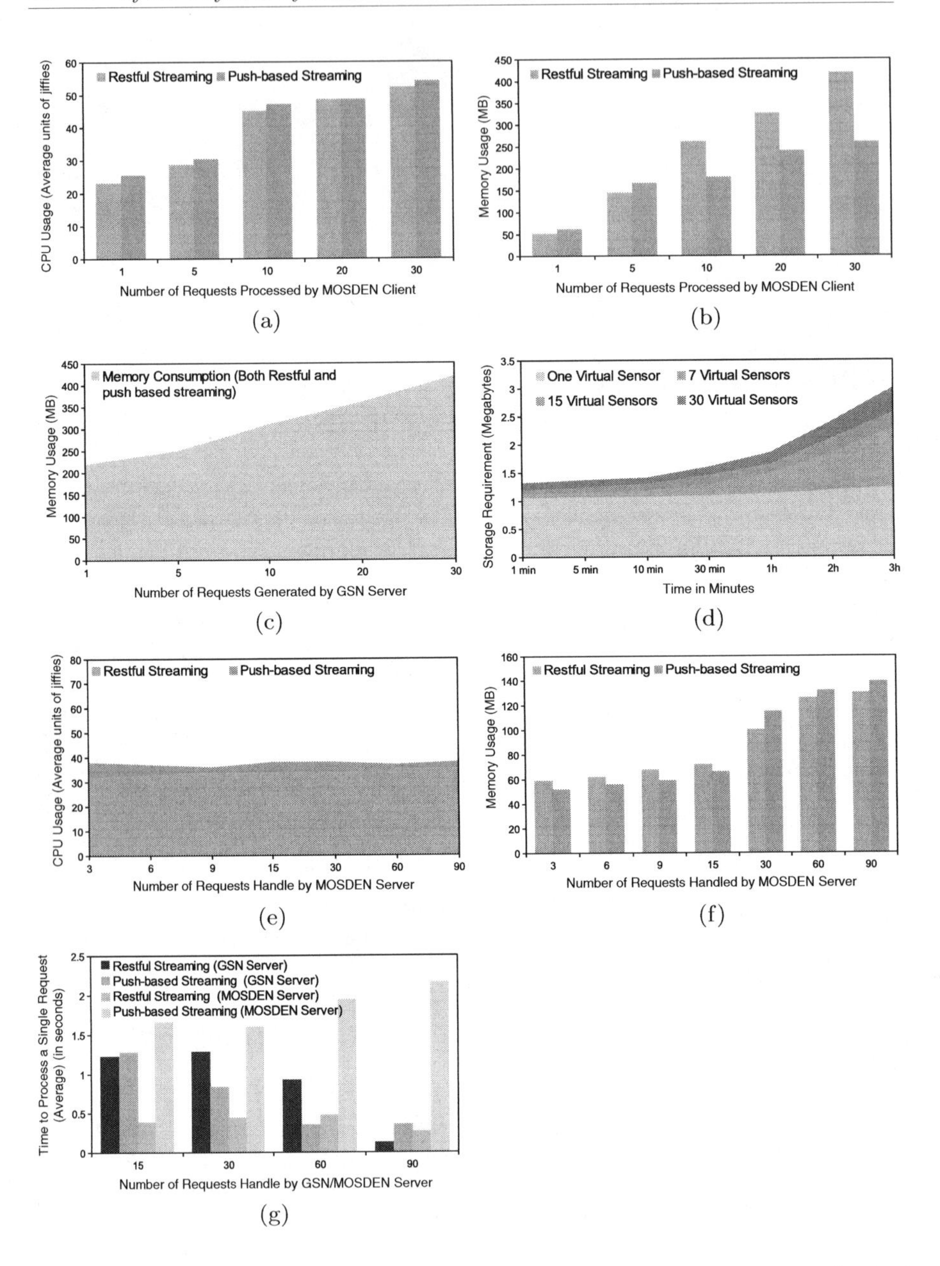

Figure 3.15: Experimental results in collaborative sensing environment.

of sensors connected to MOSDEN during the experiment in order to examine the behavior of MOSDEN from a storage requirement perspective.

According to the outcome shown in Figure 3.15(d), storage requirements are linear. It is to be noted that to stress test MOSDEN client instances, we used external sensors, on-board sensors and additional data source generators to simulate 30 virtual sensors. This further demonstrates the scalability of MOSDEN. In both GSN and MOSDEN, storage can be easily controlled by changing the history-size. History-size defines how much data record needs to be stored at a given time. Large history-sizes can be used for summarizing purposes or archival purposes. However, the amount of storage in easily predictable due to history-size, because MOSDEN always deletes old items in order to accommodate new data items. For real time reasoning history can be set to one. Considering all the above factors, it is fair to conclude that modern mobile devices have the storage capacity to store sensor data collected over long periods of time.

For the next set of experiments, we configured the testbed as illustrated in Figure 3.14(b)-Setup 2. In Figure 3.15(e) and 3.15(f), we compare the performance of restful streaming and push-based streaming techniques in terms of CPU usage and memory usage by the server mobile device (D1) which runs MOSDEN. Figure 3.15(e) illustrates the difference between CPU usage in MOSDEN when the number of requests increases.

Figure 3.15(f) illustrates the variation of memory consumption of MOSDEN when the number of requests increases. According to Figure 3.15(e) and Figure 3.15(f) push based streaming is slightly better than restful streaming. Further, it is important to note that both techniques maintain the same amount of CPU consumption over time despite the increase in requests it handles. Additionally, MOSDEN server consumes significantly less amount of memory in comparison to MOSDEN client. One reason is that MOSDEN client performs sensing activities in addition to sending data to the server. In contrast, MOSDEN server performs data requesting tasks only (from clients). As we mentioned earlier, when number of requests handled by MOSDEN increases (given that no other tasks are performed), restful streaming technique performs better in terms of both CPU consumption and memory consumption.

Figure 3.15(g), compares the amount of time (average) it takes to process a single request.[17] Time it takes to process a single request is calculated as denoted in Equation 3.1. Figure 3.15(g) shows the impact of increased overheads when using a push-based streaming technique. However, MOSDEN still processes a single request in less than a second.

$$= \frac{\text{Duration of the Experiment}}{\text{Total Number of Round Trips Completed}} \tag{3.1}$$

Overall MOSDEN performs extremely well in both server and client roles in collaborative environments. MOSDEN (as a server) was able to handle 90 requests (i.e., 180 sub requests) where each request has a sampling rate of one second. This resulted in a MOSDEN client processing 1800 data points every 1 minute and a MOSDEN server (running on a mobile device) processing 5400

[17]Time taken to process a single request is the time interval elapsed between two subsequent requests made by the server to any client irrespective of the virtual sensor.

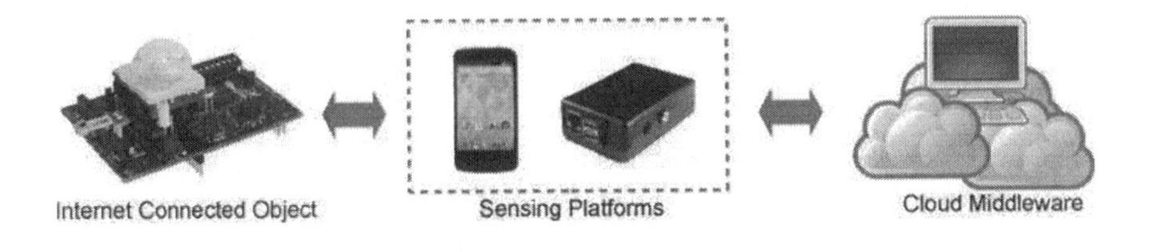

Figure 3.16: Two types of configuration are required to establish the connectivity between ICOs and cloud middleware platforms.

data points every 1 minute from distributed clients. It is to be noted, that for evaluation purposes and to validate the efficiency and scalability of MOSDEN, we conducted experiments on MOSDEN server and client under extreme loads. Such processing is intensive and rare in real-world applications. However, our experiments showed that MOSDEN can withstand such intensive loads proving to be a scalable platform for deploying large-scale *crowdsensing* applications. If MOSDEN is configured to collect data from 10 different sensors and handle 30 requests (typical of real-world situations), it can perform real-time sensing with delay of 0.4 – 1.5 seconds. When the server node is a computer (D6 as explained in Section 3.6.3) both restful streaming and push-based streaming work extremely well without visible significant differences. However, when the server node is a mobile device, which runs MOSDEN, restful streaming performs approximately 6 times better than the push-based technique.

3.7 Open Challenges and Opportunities

3.7.1 Automated Configuration

Configuration in mobile sensing middleware platforms domain can be explained in two different ways. First, configuration needs to be done between external ICOs and mobile middleware platform. Secondly, configuration needs to be performed between mobile middleware platforms and the cloud middleware. Figure 3.16 illustrates both types of configuration techniques. In order to collect data from an ICO, first it needs to establish communication with the mobile sensing middleware platform. Both of them need to negotiate common protocols and message sequences [64] in which they are capable of communicating with each other. Then, ICOs need to provide enough information (e.g., sensor ID, manufacture details, sensor types) to the mobile sensing middleware platforms, so the matching plugin can be downloaded from the cloud and configured locally. Additionally, configuration of ICO includes tasks such as configuring sampling rate, communication frequency, scheduling calendar, and data acquisition method. A similar research effort is discussed in [64]. Once mobile sensing middleware platforms collect context information from

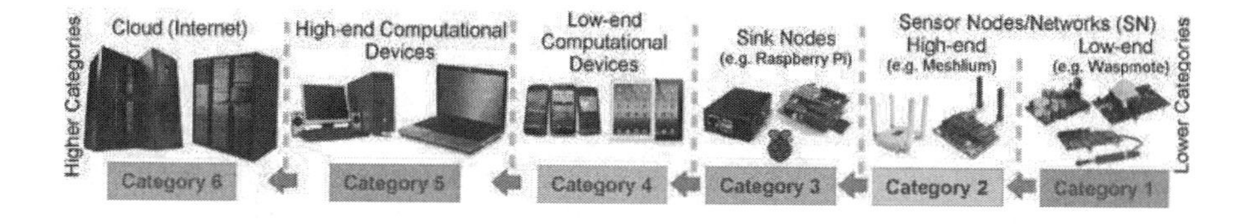

Figure 3.17: Categorization of CPS devices based on their computational capabilities. The devices belonging to each category have different capabilities depending on processing, memory, and communication. From left to right, the devices become more and more expensive; the computational capabilities also increase.

ICOs, they need to be sent to the cloud middleware so the cloud can prepare itself to accept the sensor data that may arrive later.

Additionally, such registration of ICOs on the cloud supports sensing as a service model [33]. Further, registration allows the cloud to know about sensors and data availability through each of the mobile sensing middleware platforms. Registration of mobile sensing devices includes information such as ICOs connected to it, the capabilities of ICOs, context information such as location and energy level. As a result, the cloud middleware can request the data from the correct mobile sensing middleware platform easily. Some of the preliminary research work on configuring mobile sensing middleware platforms and the cloud are in [65]. However, dynamicity and the heterogeneity of sensors increase the complexity of both configuration tasks where automation is the only viable solution.

3.7.2 Unified Middleware Platform

It is important to build a set of middleware platforms [66] that strongly integrate to each other. However, such middleware platforms need to be custom developed for different categories of devices by considering the resource limitations and capabilities of each device (e.g., processing power, memory, storage, communication bandwidth, energy availability) [67]. Some of the devices with different capabilities limitations are depicted in Figure 3.17. The MOSDEN middleware platform is built in such a way that it is fully compatible with the popular cloud-based Global Sensor Network middleware [58]. However, middleware platforms suitable for very low cost, less capable devices, such as Raspberry Pi, are yet to be developed. From the engineering point of view, difference in programming languages makes such interoperability challenging (e.g., Java/Android, .NET, Arduino, and so on). In CPS paradigm, each device has a unique role to play. Additionally, these devices should be able to perform the tasks of distributed sensing in a collaborative fashion [33].

3.7.3 Optimized Data Processing Strategy

Once the data has been collected by the sensors, it needs to be processed. However, efficiency of the entire system depends on the place where data would be processed. Let us concentrate on Figure 3.17. Once the data has been collected from the sensors on the right side, it needs to be processed at the earliest possible location, because sending data from right to left costs significant amounts of energy [62]. Some of the processing tasks (e.g., event detection using sensors attached to the same node) can be done on the sensor itself without even sending them out. In other instances, individual sensors may not have the required knowledge to process the data collected locally (e.g., event detection using sensor data collected by two or more sensors that are connected to separate sensor nodes). In such circumstances, data needs to be sent to the devices on the left. It is important to remember that the more the data moves, the more cost it adds to the overall system.

Another important factor is the time taken to process data. Sometimes, processing the real-time is critical. However, more communication increases the latency. Therefore, real-time processing should be supported by delegating processing responsibility to lower categories of devices. However, if the objective of collecting data is to process it later, then ideally sensors may cache data for some time and then push it through the devices as bulk data. This reduces the communication wastage (i.e., communication radio open and close less frequently than in real-time communication). The basic rule that governs whether to process data on the current device or not in a given situation can be found out by following the rule. Understanding the knowledge requirements is comparatively easy. However, finding out the energy requirement for data processing and communication is challenging. This challenge is yet to be addressed by the research community. The communication protocol used, energy source available to the device, and the energy consumption for processing data need to be carefully evaluated to make a decision.

3.7.4 Multi-Protocol Support

There are several popular communication protocols that have been widely used in CPS domains [68]. WiFi, Bluetooth, ZigBee, z-wave are some of them. Even though they seem few, incompatibility makes the CPS application more challenging. Each protocol has its own advantages and disadvantages. Table 3.1 summarizes basic differences in the protocols. When developing mobile middleware platforms, it is important to support all the above-mentioned protocols. Typically, data will be uploaded to the cloud using long-range protocols such as 3G or WiFi (communication between mobile middleware and cloud middleware). However, communication between mobile middleware platforms and external sensors is ideally performed through protocols such as Bluetooth and ZigBee. Additionally, it is important to address the challenge of developing a high-level framework that handles the difference in protocols behind the scenes without affecting the design of upper-level structure. So an ideal frame-

Table 3.1: Heterogeneity in Terms of Wireless Communication Technology

	ZigBee	GPRS-GSM	WiFi	Bluetooth
Standard	802.15.4		802.11b	802.15.1
System Resources	4-32KB	16MB+	1MB+	250KM+
Battery Life (days)	100-1000+	1-7	0.5-5	1-7
Network Size	2^{64}	1	32	7
Bandwidth (KB/s)	20-250	64-128+	11000	720
Transmission Range (Meters)	1-100+	1000	1-100	1-10+
Success Metrics	Reliability, power, cost	Reach, quality	Speed, flexibility	Convenience, cost

work will allow the developer of data communication in high-level (e.g., what to send when) rather than dealing with communication protocols, which one to use when and implement program code-level changes to address the differences. This will increase the efficiency and effectiveness of the CPS solutions but also save a significant amount of development time.

3.7.5 Modular Reasoning, Fusing, and Filtering

In the current state of the MOSDEN middleware, we only support limited SQL query-like filtering and fusing (e.g., average). It is important to build a plugin architecture that is similar to the one we introduce to collect data from external sensors, to support modular reasoning (e.g., secondary context generation using primary context). A preliminary work that addresses this challenge is *Dynamix* [69]. *Dynamix* is a plug-and-play context framework for Android. *Dynamix* automatically discovers, downloads, and installs the plugins needed for a given context sensing task. *Dynamix* is a stand alone application, and it tries to understand new environments by using pluggable context discovery and reasoning mechanisms. It does not provide a server-level solution. Context discovery is the main functionality in *Dynamix*. In contrast, MOSDEN is focused on allowing an easy way to connect sensors to applications in order to support sensing as a service model in CPS domain. We employ a pluggable architecture that is similar to the approach used in *Dynamix*, in order to increase the scalability and rapid extension development by 3rd party developers. Activity recognition is one of the popular reasoning tasks performed in mobiles. In plugin based architecture, it is critical to have a cloud companion integrated to the solutions so the big picture can be understood by the cloud and then configure the mobile sensing middleware platforms as necessary, because mobile sensing platforms have only limited knowledge about the resources they have access to, in comparison to cloud middleware.

3.8 Summary

The number of mobile devices connected to the Internet is growing at a rapid pace. A significant portion of these devices is mobile devices today. However, it is expected that billions of different types of resource constrained computational devices will be connected to the Internet over the coming decade. On the other hand, the number of sensors deployed around us is increasing. It is an increasingly important task to collect data from these sensors in order to analyze and act upon it. In this chapter, we described our approach on building a mobile middleware platform, called Mobile Sensor Data Engine (MOSDEN), for resource constrained devices. MOSDEN can be installed in mobile devices (smartphones and tablets) and can be used to collect data from both internal and external sensors. Due to the plugin architecture, MOSDEN can retrieve data from virtually any smart device. Further, it has the capability to perform limited data processing and filtering tasks.

We also discuss a number of different ways that existing mobile system platforms have been used to support sensor data collection and processing. This chapter comprises a number of performance evaluation results where MOSDEN has been tested in stand-alone mode as well as in collaborative mode. In the later part of this chapter, we presented different real-world applications for which MOSDEN can be used. Finally, it is important to mention that the goal of building a unified middleware platform that supports a broad range of devices, ranging from low-level sensors to smartphones to personal computers to the cloud, is yet to be achieved by the research community. Addressing the open challenges mentioned previously will help to move in that direction.

Acknowledgment: Authors acknowledge support from SSN TCP, CSIRO, Australia and ICT Project, which is co-funded by the European Commission under the seventh framework program, contract number FP7-ICT-2011-7-287305-OpenIoT. The authors also acknowledge help and contributions from The Australian National University.

Chapter 4

Naming, Addressing, and Profile Services for CPS

Chi Harold Liu

Beijing Institute of Technology, China

CONTENTS

The preceding chapter discusses mobile sensing platforms and their applications, and this chapter presents another important factor in the architecture of CPS: Naming, Addressing, and Profile Services. Given massive amounts of heterogeneous devices deployed across different platforms in cyber physical systems, how to provide efficient upstream sensory data collection, content-based data filtering/matching, and efficient downstream control is a common issue. We will discuss a middleware component to service dynamic application needs, and sensors/actuators deployment and configurations across different platforms.

4.1 Introduction

In the field of CPS, the lack of a *de facto* standard architecting a naming, addressing, and profile server (NAPS), as a middleware [70] interoperable with heterogenous platforms has become a key limiting issue on its proliferation to deployment [71]. The research community is hearing the strong desire from application developers to avoid learning heterogeneous communication/networking protocols in use, but be provided a homogeneous naming and addressing convention, so that they are able to retrieve the data from sensors and control the actuators of different platforms and network domains. Toward this end, a higher layer of device naming-addressing mapping should be provided to integrate with legacy systems and different platforms. As for device naming, the convention should contain key elements of device meta-data, such as device type and domain information; while for addressing, its format allows the granularity of efficient accessibility and addressability to the physical world. Profile services are also needed to aid the application query and system configurations, like device status and presence. Furthermore, sensing tasks are always achieved by a group of devices with similar sensing capabilities, and thus NAPS should provide device group management functionalities, such as to create, update, read, and delete (CURD) groups (and its tree-structured subgroups). In this way, application development logic is greatly simplified where only a device group name is needed and NAPS handles the internal mapping. As a middleware, it should extend its usability by providing abundant external interfaces.

IPv4, IPv6, and Dynamic Name Service (DNS) are usually considered as the candidate standards for naming and addressing; however due to the lack of communication and processing capabilities of many small and cheap devices (like RFID tags) it is quite challenging to connect every "thing" with an IP. Furthermore, with the increasing number of end devices, even IPv6's address space may not be enough. On the other hand, industry standards have put much effort into each application domain. EPCglobal [72] uses a 96-bit binary sequence to identify each RFID tag, and the object name service (ONS) for URL translation. OPC-UA [73] defines client-server based models for industrial production line solutions, where an abstract address space is formed by a mesh topology. In it, each node represents a sensor in the production stage and the edge between two nodes represents the stage-by-stage relationship during the production. As an overall service architecture, ETSI [74] proposed a solution interworking with 3GPP machine type communication (MTC) standard [75], to support machine-to-machine (M2M) communications when upgrading from traditional cellular networks where each device is with a unique international mobile subscriber identity (IMSI) and is IP addressable. Furthermore, as a service layer architecture, it defines a variety of service capabilities (SCs) including a Network Reachability, Addressing, and Repository (NRAR) SC. However, it has no technical details thus far. Our goal in this work is to work with any service platform as a middleware at the back-end data center. Therefore, all these efforts pay attention only to a specific network or application domain, instead of a *common platform* managing different technologies and standards.

4.1.1 Scope and Assumptions

There is no common platform interoperable with different platforms to hide this heterogeneity and provide a transparent naming service to applications. We therefore designed a CPS - application infrastructure (CPS-AI) and its management platform (out of the scope of this chapter). The key technical entablements of CPS-AI are: application gateway (AG), NAPS and its service registration portal (Portal), and real-time operational database (RODB). AG coordinates the data filtering and processing, and controls message delivery based on a uniform device naming and addressing convention in NAPS. The goal is to have applications access devices across different platforms without knowing their languages in detail, but focusing on the development logic only. The position of NAPS extends the functionality comparable to DNS in the Internet, to the profile services such as storage and query. We next present three assumptions of this work.

First is service discovery. Since the scope of NAPS is a middleware component at the back-end data center to hide the heterogeneous protocols and standards, here we assume that service discovery has already been successfully performed by each platform individually, and stored in our NAPS repository. Examples are service discovery server enhanced from ETSI M2M service architecture by InterDigital [31], discovery service set in OPC-UA standard, and protocols like Universal Plug and Play (UPnP) [76], etc.

Second is the authentication, authorization, and accounting (AAA). Although it is not the focus of this work, the design can largely leverage the Network Security Capability (NSEC) SC in ETSI M2M service architecture. It uses a key hierarchy, composed of root key, service key, and application keys. Root key is used to derive service keys through authentication, and key agreement between the device or gateway and the M2M SCs at the M2M Core. The application key, derived from service key, is unique as per M2M application. Issues like distributed denial-of-service (DDoS) attack will be discussed in Section 4.7.

Finally, we assume that wireless imperfection like packet errors and interference have been handled by the communication stack of each access network. Solutions from PHY layer techniques (e.g., antenna techniques, modulation and coding) and MAC/network layer protocols (e.g., scheduling and routing) are a few examples. Therefore, any wireless issues are completely transparent to the service layer operations, or the NAPS middleware considered in this chapter.

4.1.2 Contributions and Chapter Organization

Our contributions are summarized as fourfold. First, we propose a complete and detailed design of NAPS, including its key system flows, interfaces, and individual module designs. Second, we propose a unique device naming and addressing convention interworking with different platforms, and we show its applicability to a few widely used standards and protocols. Third, we propose an efficient identifier generation scheme, not only used during data transportation, but also to facilitate the data filtering and matching. Fourth, we provide CURD operations on device, device type, and device group profiles, in the RESTful design style [77] over HTTP at runtime. Meanwhile, we provide a JDBC library so that external applications can access large amounts of profile information (in the scale of gigabytes), where HTTP falls short. Finally, we demonstrate its usage by a case study in a smart building environment and show performance evaluation results on system throughput. To the best of our knowledge, this is the first piece of work to tackle the fundamental design issues of naming, addressing, and profile services as a middleware across different platforms by demonstrating its usage through a real implementation. A summary of important acronyms used in this chapter is listed in Table 4.1.

The rest of this chapter is organized as follows. Related research activities are introduced in Section 4.2. System context and flows are described in Section 4.3. Section 4.4 presents the detailed system design and implementation issues. Next, performance evaluation is given in Section 4.6. After discussing a few related practical issues in Section 4.7, a conclusion is drawn in Section 10.7.

Table 4.1: Summary of Important Acronyms

Acronyms	meaning
M2M	Machine-to-machine
SC	Service capability
NAPS	Naming, addressing, and profile server
RFID	Radio frequency identification
CURD	Create, update, read and delete
NRAR	Network reachability, addressing and repository
OPC-UA	OLE for process control - Unified Architecture
CPS-AI	Cyber Physical Systems application infrastructure
RODB	Real-time operational database
AG-FEP	Application gateway - front-end processor
AG-BE	Application gateway - back-end
AG-CC	Application gateway - command controller
AAA	Authentication, authorization, and accounting
IBM WAS-CE	IBM Websphere Application Server - Community Edition
devID	Device identifier
Java PO	Java persistent object
Java DAO	Java Data Access Object
API	Application Programming Interface
URI	Uniform resource identifier
XSD	XML Schema Definition
JDBC	Java Data Base Connectivity
JVM	Java virtual machine

4.2 Related Work

EU FP7 project CPS-Architecture (CPS-A) extensively discussed the existing architectures, protocols, and platforms [78]. Besides, in [79] the authors propose a framework to interconnect sensors running 6LoWPAN [80], where IEEE 802.15.4 and IPv6 were considered to connect wireless devices to the Internet. In [81], they provide an XML schema to encode device profile information including its local name. Web-of-things (WoT, [82]) makes use of popular Web languages for building applications involving smart things and users.

As for industrial standards, based on DNS, the pure IP solution [83] is favored due to the recent development of IPv6 to connect "things" for CPS. EPCglobal [72] specializes in the use of RFID in the information rich and trading networks, especially for logistics. In it, similar to DNS, an object name service (ONS) is designed to translate a 96-bit binary sequence to a URI, which directs the query to a (set of) database(s) called EPC information services for information retrieval and update. OPC-UA [73] defines an address space where devices are interconnected to form a mesh topology. The connectivity represents the production line sequence, where the directional edge called "reference" links the next stage of behavior. ETSI M2M service architecture [74] assumes each M2M device is IP addressable. In 3GPP MTC [75], they propose to use the IMSI as the internal identifier for signalling and charging, while providing external identifiers to include domain information under the control of a network operator, and flexibly allocated descriptions as the customer friendly local identifier.

On the other hand, naming and addressing for wireless sensor networks have been extensively investigated [84, 85, 86, 87]. The first kinds of approaches rely on the efficient address allocation among nodes, where in [88, 89] the assigned addresses are reused spatially and represented by variable length of codewords. This was later extended in [90] which used the prefix-free Huffman encoding of node addresses based on the energy map, where nodes with little battery life left will have the advantage of a short address and check period. The authors of [91] proposed the "attributed-based naming." In [92], clients use an intentional name to request a service without explicitly listing the end-node that ultimately serves the request. This level of indirection allows applications to seamlessly continue communicating with end-nodes even though the mapping from name to end-node addresses may change during the session. Finally, a unique identifier generation problem is investigated in [93], where they designed a correlated lookup scheme with a distributed hash table to improve the performance of identifier management.

However, none of these schemes is motivated from the service layer as part of the middleware at the back-end data center, nor do they tackle the fundamental problem of providing a homogeneous, both human and machine understandable, and unique *naming and addressing convention* across different platforms. Furthermore, none of them supports profile services and legacy system integration.

4.3 System Flows

To support the high scalability requirement, we further decompose AG into four modules, front-end processor (FEP, for data collection and format transformation), command controller (CC, for application command parsing and translation from NAPS), back-end processor (BE, for rule-based data filtering and matching), and message queue (MQ, for publish-subscribe based topic services [94]). Then, users can identify its bottleneck and scale up/out the corresponding component; for example, to deploy MQs for a large number of applications in a cloud. The list of components that interact with NAPS either offline or at runtime are AG-FEP, AG-CC, AG-BE, applications, portal, and RODB, as shown in Figure 4.1 for system context and associated interfaces. It is worth noting that the overall CPS-AI platform is only used as an example to demonstrate the system flow of NAPS, whereas its applicability can extend to any external component with similar interfaces. We next present three key system flows, service registration and configurations, upstream data collection and downstream command delivery. In all aspects, an AAA server

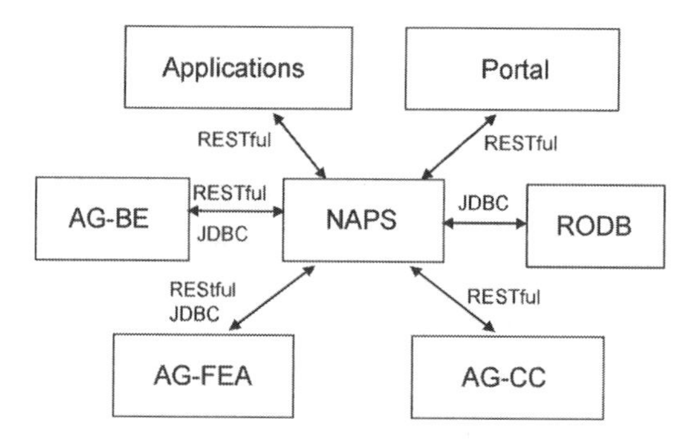

Figure 4.1: System context of NAPS and the associated application-layer interfaces.

interacts with NAPS for security authentication and authorization, and we left the details for future work.

4.3.1 Device Registration and Configurations

As discussed earlier, service discovery is performed at an individual platform beneath the data center service layer, and NAPS only provides a set of interfaces to facilitate the device registration, either automatic or offline. The registered capabilities include the ones offered by devices, device types, and device groups, and thus the repository stores the corresponding profile information. The provided interfaces are based on the RESTful design style where standard HTTP request/response is used to transport the data. It is worth noting that before the response is returned to the client, we generate a unique device identifier, or "devID" for the rest of the chapter (see Section 4.4.3). It contains key elements of the device meta-data. Note that this devID generation process is also applicable when device type and device group are registered. For service bootstrap, other components like AG and RODB can establish JDBC connection to retrieve these meta-data from the database.

4.3.2 Upstream Data Collection

As shown in Figure 4.2(a), when AG-BE receives the raw data from a CPS platform, it translates the devID to the corresponding device name (devName) as more friendly to the application. Meanwhile, AG-BE makes use of devID to perform efficient content-based data filtering and matching. For example, one application configures a topic on MQ to aggregate and average the room humidity data in a smart building environment. Then, since the designed devID contains key elements of device meta-data such as associated domain

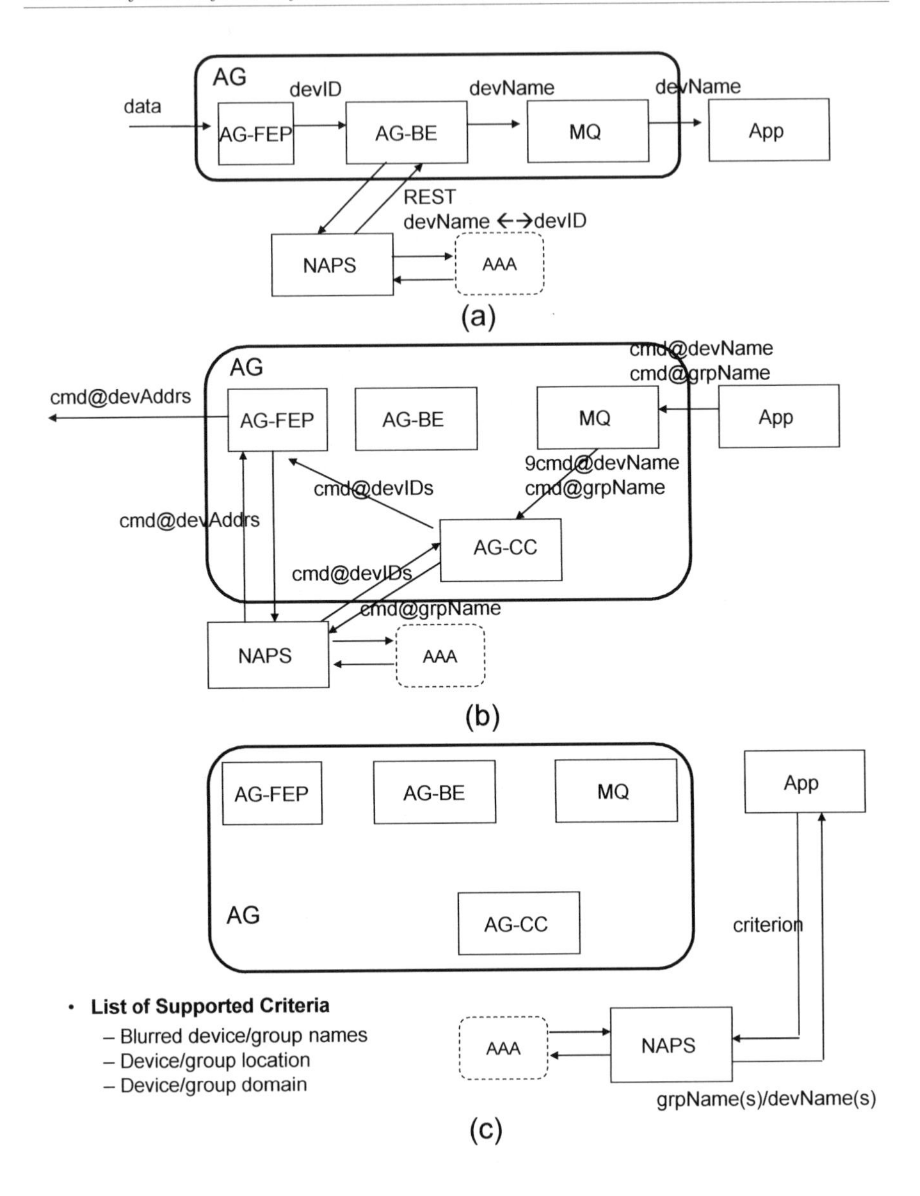

Figure 4.2: System flow: (a) upstream data collection, (b) downstream command delivery, and (c) application query.

information and device type, it helps to categorize, filter and select the exact set of raw data from the massive data pool. This can be achieved by masking certain encoded profile parameters. To save the overhead, we use devID as the unique identifier in the transport (e.g., through the carrier public network and

CPS private network as shown in Figure 2.1), while using devName containing a set of human-readable properties for applications.

4.3.3 Downstream Command Delivery

Figure 4.2(b) shows the downstream system flow when control messages are initialized by the application to specific device groups. First, device (group) name is passed to the AG-CC, and the latter retrieves the (list of) devID(s) from NAPS over the RESTful interface. Then, AG-FEP translates the devID(s) to the corresponding device address(es) from NAPS, in which it specifies how to address the command(s) to the exact (group of) device(s). In this way, similar to the functionality of DNS in the Internet, NAPS performs the name-to-address resolution. It is worth noting that our aim is not to design completely new transport protocols to inter-operate with heterogeneous network domains like cellular networks and ZigBee. However, we only define a thin layer for address resolution, as a uniform convention to unify and coordinate different platforms. In our addressing convention (see Section 4.4.2), it specifies the way to route the command to the Internet gateway, which is IP addressable, such as the M2M gateway in ETSI M2M service architecture, OPC-UA server, or WiMax base station (as considered in the case study). In a hierarchy, these gateways further route the command to the next level gateway (e.g., the ZigBee coordinator) which maintains its own addressing mechanism to the device.

4.3.4 Application Query

Application developers may be the same as or, as a future trend, different from the device owners. Therefore, their development logic entirely relies on identifying the right set of devices from the physical world, which may eventually belong to different network domains/platforms. Toward this end, since the device, device type, and device group profiles are registered and stored in NAPS repository, we allow search services to retrieve a list of devices and device groups with certain geographical, domain, and device name information. Furthermore, this process can be coupled with downstream command delivery procedures where a retrieved list of device and device group names is used to issued commands to the physical world.

4.3.5 Integration with Different CPS Platforms

Nearly all third-party device vendors and platform operators have their own naming mechanism, and it is likely that they also have already developed their own applications. Therefore, to ease the integration process, the exposure of the proposed uniform device naming is not the most convenient way. Toward this end, NAPS provides the translation between our uniform naming convention and the legacy naming, offering shared services to different vendors and applications. We thus provide two RESTful interfaces, `getDevOldNamebyDevName()` and `getDevNamebyDevOldName()`, where the former is used when the upstream data are received by the application attached with our device name, and thus to be translated to the legacy name, and the latter is used when applications issue commands to the actuators by translating the legacy naming to the proposed one. In this way, existing naming platforms like 3GPP MTC and EPCglobal can be seamlessly integrated with our NAPS middleware.

Since one of the functionalities of NAPS is to store the registered profile information, among many choices, our implementation uses IBM DB2 as the relational database to store all devices, device types, and device groups related profile information. A JDBC library is used as an interface for the Web container (where Java code resides) to access the database (see Figure 4.3). We choose IBM WAS-CE as this container that hosts an Apache Tomcat application server and other necessary programming supports. Within the Web container, NAPS implements the data persistence, data access object (DAO),

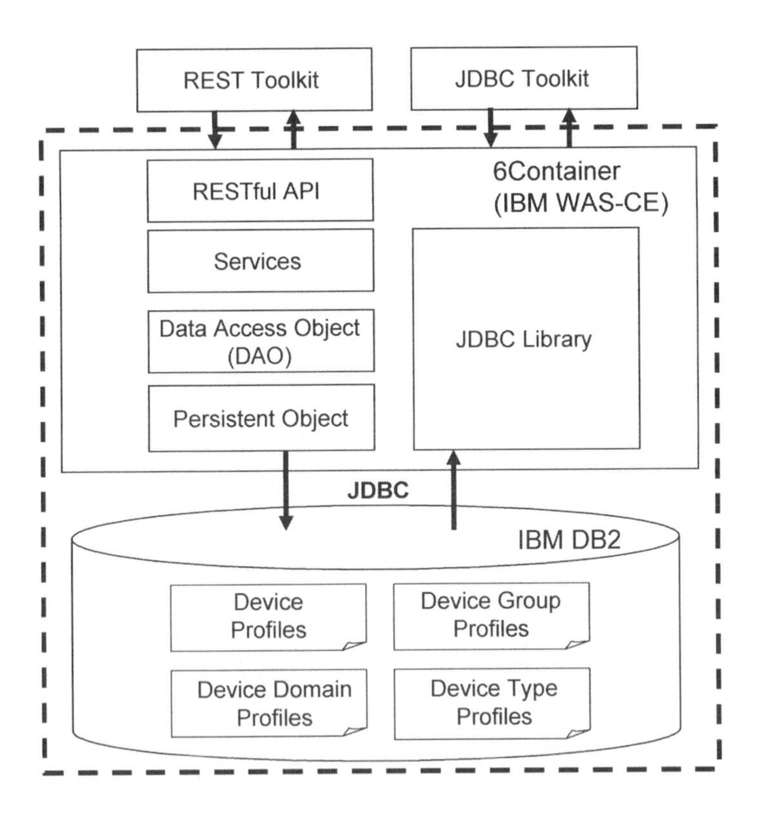

Figure 4.3: NAPS component design.

service layer, and RESTful APIs; and in parallel, JDBC library and client-side toolkits. Next, we show key functionalities of each internal module.

4.4 System Designs and Implementations

Java PO: Java persistent object, which specifies the one-one correspondence from a Java object to the database entries, columns, and relations. Persistency is also implemented by using the open-source packages like Apache openJPA.

DAO: basic database operations, like delete, get, query, save, update, and above operations in batch are defined and implemented by using the Java PO.

Services: in this layer, RESTful resource URIs are decomposed into the Java PO consistent with the same database design, e.g., entry names, column names, and their corresponding values. Then, HTTP method calls are translated into internal APIs which will be further decomposed into database basic operations defined in the DAO layer. Service logics are also maintained, and in particular for device deletion related operations, this layer will help check the related device group profile and perform the update accordingly.

JDBC library: provides a wrapper from the JDBC operations to the Java package.

RESTful API: serves as the external interface to send and/or receive response/request to the clients.

For the sake of completeness, we also describe table designs on IBM DB2, as shown in Figure 4.4. Four key tables are device, device type, device group, and device domain tables, with the one-to-many mapping between the device types and devices, many-to-many mapping between devices and device groups, and one-to-many mapping between device domain and devices.

4.4.1 RESTful Interfaces

We expose RESTful APIs to external users. In REST, clients initiate requests to servers that process requests and return appropriate responses. Neither the client nor the server needs to store the transitional states between the exchanged messages. This restriction isolates the client of changes in the server side. REST uses a set of HTTP methods, known as POST, GET, PUT, and DELETE, corresponding to a complete set of CURD operations. To this end, this is a perfect technology to describe application activities including profile browsing, look-up, update, and delete.

The requests and responses in REST are built around the transfer of "representations" of "resources." We define resources in NAPS as the device, device type, and device group meta-data, rather than the "actions" on it. For instance, to retrieve an entire device record, one should use the GET method requesting the URI as: `.../devices/devid/{id}`, where the `devid` denotes the devID as the key to the device profile table entry. Figure 4.5 shows the

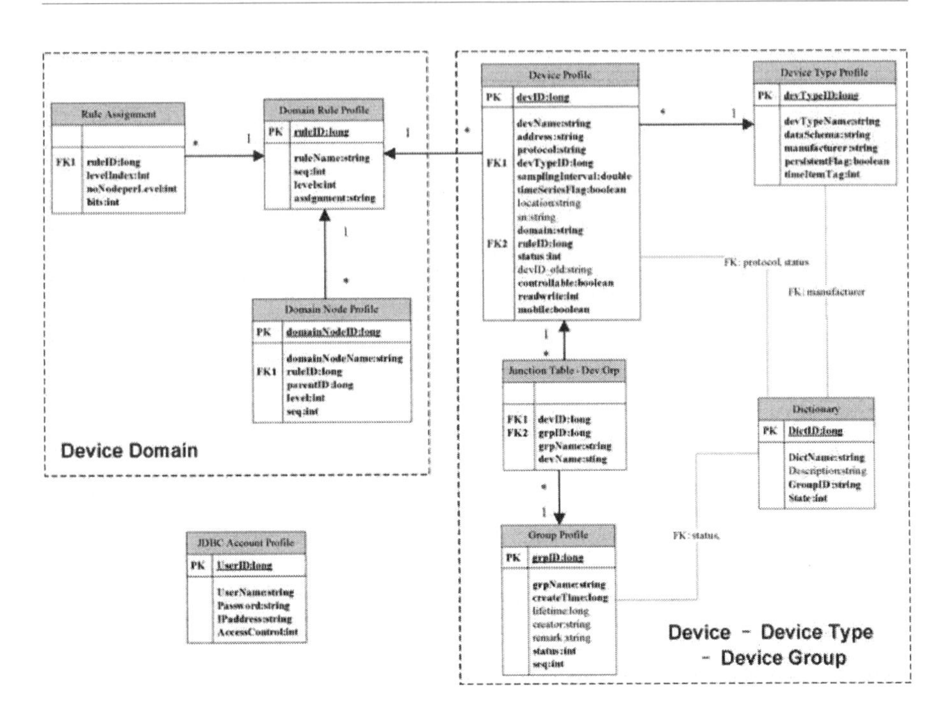

Figure 4.4: Table designs for device, device type, device group, and device domain tables.

proposed tree-based URI structure. Followed by the NAPS IP address and port,[1] three types of *root* resources are defined, i.e., devices, device types, and device groups. After, we use the format of attribute name-value to explicitly indicate a specific *leaf* resource as the operand. For example, if one writes "`...devices/devid/{724487363}`", then it maps to the record entry with its key equals to `devid` and value equals to 724487363. However, if detailed attribute name-value information is omitted, like "`...devices/`", "`...devices/*`" or "`...devices/devid/*`" , then it operates on all leaf resources (all devices in this case). Note that our design also supports the multiple leaf resource operations under one root resource, i.e., we use "&" to concatenate multiple values, e.g., "`.../devices/devid/{id1}&{id2}&{id3}`".

We next explain the use of four HTTP methods on our RESTful resource.

POST: to submit the data from the client to be processed at a target resource, resulted in the creation of a new or an update of the existing resources.

[1] Or server URL; and if considering the cluster deployment, this address corresponds to the address of server dispatcher, allowing to direct the client request to the corresponding NAPS.

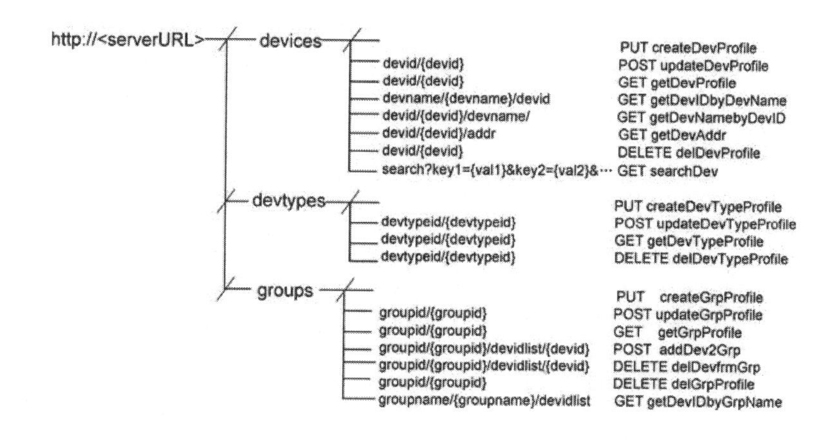

Figure 4.5: RESTful URI design for NAPS.

The data are included in the body of the request, in the format of XML. We use this method to register a new profile.

GET: to request the information from the specified resource, e.g., making a request GET in `.../devices/devid/{724487363}` will return as a response to the device record with devID=10. We use this method to retrieve the profile.

PUT: updates the resource's particular set of information, and if request URI does not exist, it allows to create a new resource. If one aims to update the serial number of a device, clients make a PUT request to: `.../devices/devid/{724487363}/serialnumber/{456WEFGA}`, where 456WEFGA is the refreshed data. We use this to update an existing profile.

DELETE: to delete a specified resource without an XML body. For instance, if one requests URI: `.../devices/devid/{724487363}`, it deletes the device record with devID=724487363, however the service layer will check its relations to the device group to make sure the mapping is also removed. We use this method to delete an existing profile.

Tables 4.2 and 4.3 show RESTful APIs to retrieve and search devices (under condition), respectively.

4.4.2 Naming and Addressing Convention

We propose a novel naming convention for devices and device groups across different platforms in a form of:

dev://domain-series/devtype/legacy-name,

grp://domain-series/target/policy/grp-name,

where the prefix distinguishes its category (as for devices or device groups), followed by a series of device domain information. The device domain is

Table 4.2: RESTful API to Get Device Profile Information

URI	http://<serverURL>/devices/devid/{724487363}&... /paging?startindex=0&count=100 http://<serverURL>/devices/devid/* /paging?startindex=0&count=100
Method	GET
Query string	startindex: start index, default=0 count: No. of records needed., default=50
Returns	200 OK & XML response 404 Not Found 500 Internal Error

Table 4.3: RESTful API to Search Devices under Condition

URI	http://<serverURL>/search?key={val1}&key={val2}&...
Method	GET
Query string	key=location, domain, devname, devtype
Returns	200 OK & XML response 404 Not Found 500 Internal Error

organized in a tree structure, written in the above naming convention back trace from the leaf to the root node of the domain tree. After the domain series, for devices we use the device type information to further categorize all devices associated with a domain node on the leaf of the tree, and finally the legacy naming (e.g., serial number) from the production phase. Meanwhile, device groups use the monitoring target (e.g., the room temperature) and grouping policy for detailed classification.

Device domain information refers to either their deployed geographical information, or logical organization of these devices. In NAPS, we allow to store multiple device domains, and each domain indexed r is a tree structure with depth d_r and width of each level as $\{w_r^i, \forall i = 1, \ldots, d_r\}$. In other words, domain r is composed of total $N_r = \sum_{i=1}^{d_r} w_r^i$ domain nodes. We call the "partition" of a domain tree by parameters $(d_r, \{w_r^i\})$ the "domain rule," and the corresponding data structure representation "rule assignment." For example, the rule assignment parameters are the number of bits to store the domain nodes in each level of the tree. The meta-data of each node include the name, its parents and children, and other properties; see Figure 4.4. In practice, project managers of a CPS system will carefully plan the device

Table 4.4: Examples of Device Addressing Convention with Different Protocols

Protocol	Example
OPC-UA	nodeID@opcua/<OPCUA-ServerURL>
ZigBee	short-address@zigbee/<coordinatorIP:port>
RFID	binary-epc@rfid/<readerIP:port>
IP-based	<ip-address:port>@http
Bluetooth	MAC-address@bluetooth/<receiverIP>
WiFi	192.168.x.x@wifi/<access-point-IP>
USB	VendorID-ProductID@USB/<ip-addr>

deployment at a site of interest, where the first step is to plan an overall device domain. For example, in a smart building environment, how many temperature and humidity sensors of what device type should be deployed at which location, and this deployment stage eventually specifies a domain tree structure, and is stored in NAPS repository.

As for device addressing convention, we propose to use the format:

$$\underbrace{\text{address-1@protocol-1/.../address-}n\text{@protocol-}n}_{\text{non-IP networks}}\text{/IP-address,}$$

to accommodate heterogeneous protocols and standards in use across different platforms. To allow the granularity of addressing the device in a hierarchy, we repeat the element "address@protocol" one after the other from the device to the Internet gateway, which is IP addressable (as the last part of the convention). Table 4.4 shows a few examples for well-known protocols. An example of naming a device with our address convention is presented in Section 10.6.

4.4.3 Generating the devID

As mentioned earlier, when device profile information is registered either manually or automatically from each CPS platform, a devID is automatically generated. We propose to use a 64-bit long integer. It is used when the data and control messages transfer between the access network, carrier public network and CPS private network, to save the communication overhead over any character-based naming. However, one cannot randomly generate this sequence to avoid potential collisions when a cluster of NAPS is deployed in a cloud environment. Meanwhile, the allocation of these 64 bits should have to support other components. For instance, AG will deploy its policy-based rule engine like MQ topic according to the device domain information, where one example is that applications may efficiently query all devices with mobility

Table 4.5: Bit Allocation for devID (in total 64 bits)

ID category	device type	control flag	r/w flag	mobile flag	domain rule	domain series	sequence
5bits	6bits	1bit	2bits	1bit	5bits	34bits	10bits

for their connectivity and presence at runtime by simply masking a portion of this devID. Toward this end, we propose the following design, as shown in Table 4.5. The category field identifies the type of sequence in the database (for devID this field equals to 1), followed by the device type, controllable flag, read/write access flag, and mobility indicator. The last part is a series of domain information, starting from the domain rule sequence r and back trace from the leaf to the root domain nodes of that domain rule. A detailed example will be presented in the next section.

4.5 A Case Study

4.5.1 *Device Deployment, Naming, and Addressing Format*

In this section, we demonstrate the use of NAPS in a real smart building scenario. As shown in Figure 4.6, we deploy three temperature sensors, one humidity sensor, one alarm, and one camera, all wirelessly connected by ZigBee with a central ZigBee coordinator. All devices are deployed in room 308 at Floor 3 of IBM Research in Beijing, China. In particular, three temperature sensors form a multi-hop wireless network. Then, in order to assist the device naming, we first define the device domain as shown in Figure 4.7. This phase is usually associated with the initial planning stage where project managers decide the location, device type, and number of sensors for deployment. In this example, we use sensors' deployed geographical information to form the device domain tree. Without loss of generality, we denote the domain rule index $r = 1$ with depth $d_{r=1} = 3$ and width of three levels as $w^1_{r=1} = 1, w^2_{r=1} = 3, w^3_{r=1} = 6$, respectively. Collectively, we have $N_{r=1} = 10$ domain nodes in total. Then, domain rule assignment tells the number of bits used to encode these domain nodes, as 1 bit, 2 bits, and 3 bits for three layers, respectively. Based on this, we sequentially allocate bits to each domain node, as used later for devID generation of each sensor, as shown in the figure. Besides, we also register the device type information in NAPS repository, and use encoding (010101)B, (010110)B, (010111)B, and (011000)B to represent temperature, humidity, alarm, and camera, respectively.

Based on the proposed naming convention and devID generation scheme in Sections 4.4.2 and 4.4.3, Table 4.6 summarizes the result. Take temperature sensor 1 for example; its generated devID is 0xAA450000000F001, where ID

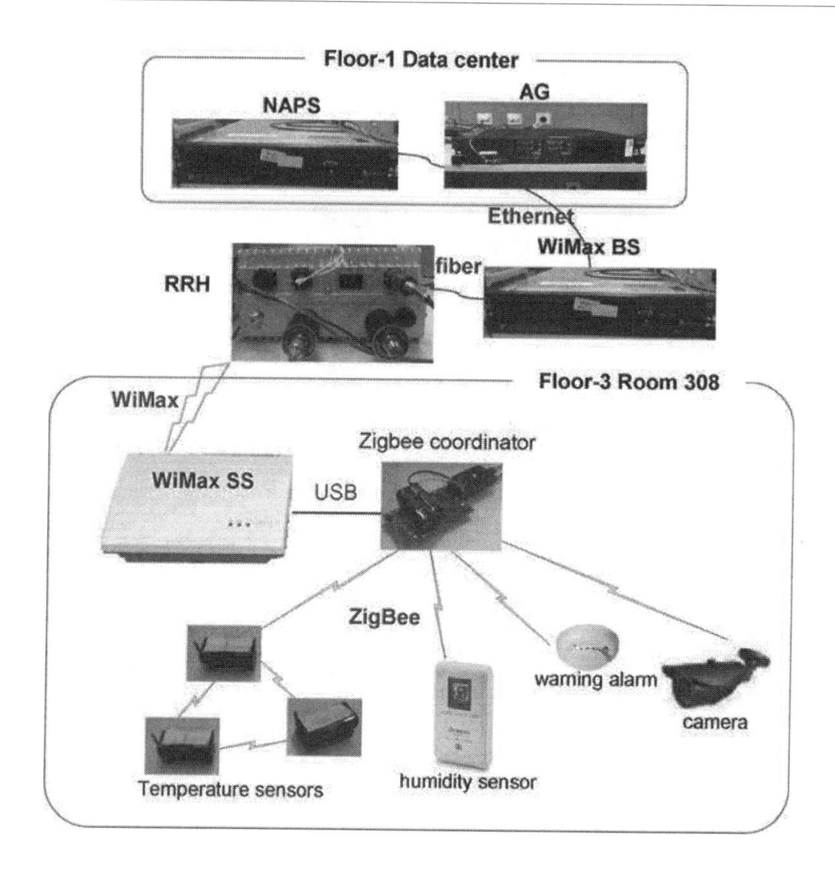

Figure 4.6: Device deployment in a smart building environment.

category=(00001)B, device type "temperature"=(010101)B, control flag "controllable"=(0)B, r/w flag "read only"=(01)B, mobility flag "static"=(0)B, domain rule "IBM site"=(00101)B, domain series = $(\underbrace{0\ldots0}_{28}\ 111100)$B[2], and "sequence"=(1)B. Furthermore, users may create a device group to monitor the "average" room temperature on a daily basis, then we generate its naming as: `grp://rm308.floor3.IBM/temperature/daily-average/group1`. To do this, a user deploys a rule in the AG, and simply masks 0xAA000000000F000 with the devID of the collected data, or $(1\underbrace{010101}\underbrace{0\ldots0}_{37}\underbrace{1}_{\text{IBM}}\underbrace{11}_{\text{Floor3}}\underbrace{100}_{\text{room308}}\underbrace{0\ldots0}_{10})$B.

Behind the access network (ZigBee based sensor network in this case), we deploy a WiMax subscriber station (SS) in the same room that connects the ZigBee coordinator by a USB cable. At Floor 1 of the building, we house a

[2]We assign domain node IBM, Floor 3, and room 308 as (1)B, (11)B, and (100)B, respectively, see Figure 4.7. Meanwhile, the upper-most 28 bits are filled by zeros.

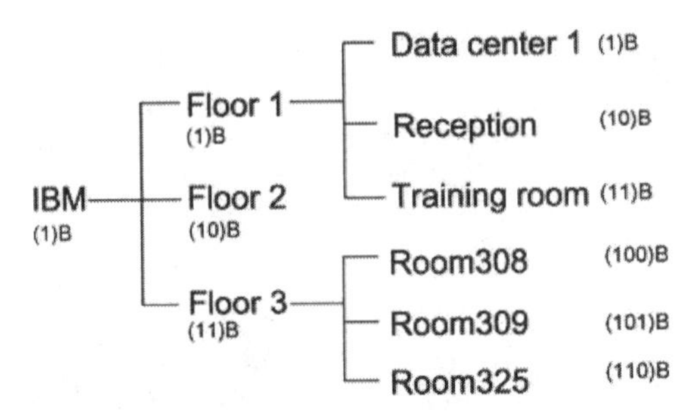

Figure 4.7: Device domain structure for the considered smart building scenario.

Table 4.6: Device Naming Format and devID of Six ZigBee Sensors

device	devName begin with dev://rm308.floor3.IBM	devID
temp. sensor 1	.../temperature/DER452SA	0xAA450000000F001
temp. sensor 2	.../temperature/LDKE4512	0xAA450000000F002
temp. sensor 3	.../temperature/235S4FDE	0xAA450000000F003
humidity sensor	.../humidity/98SWLK12	0xAC450000000F001
warning alarm	.../alarm/273FDS43	0xAFC50000000F001
camera	.../camera/091ASEKL	0xB1C50000000F001

small but compact data center comprised of an AG (on an IBM Workload Deployer) and a NAPS server (our Java code is running on an IBM System x3650). AG connects to the WiMax base station (BS), which is implemented as a software-defined radio system on hardware IBM System x3650. By using a remote radio head (RRH), BS extends the communication range to cover the entire building. The communication channel between SS and AG is the standard WiMax air interface. WiMax employs a connection oriented MAC layer and each connection is identified with a set of 16-bit connection identifiers (CIDs). Upon entering into networks, the SS is assigned with three CIDs in each direction, i.e., the basic CID used for initial registration derived from its 48-bit MAC address, primary management CID used for connection management and authentication, and secondary management CID used for management messages such as DHCP. In addition, each terminal is allocated with transport CIDs for data services. Upon service discovery and registration, SS sends the list of capabilities to AG in the REG-REQ message, and then

Table 4.7: Device Addressing Format of Six ZigBee Sensors

device	addressing
temp. sensor 1	0x4FE2@zigbee/03F0-2D12@USB/9.186.1.133
temp. sensor 2	0x4FE3@zigbee/03F0-2D12@USB/9.186.1.133
temp. sensor 3	0x4FE4@zigbee/03F0-2D12@USB/9.186.1.133
humidity sensor	0xD3E1@zigbee/03F0-2D12@USB/9.186.1.133
warning alarm	0xF2A1@zigbee/03F0-2D12@USB/9.186.1.133
camera	0x82AD@zigbee/03F0-2D12@USB/9.186.1.133

AG responds with a REG-RSP message along with supported capabilities. Finally, SS will acknowledges REG-RSP with ACK.

As for addressing, SS and AG are provided with a public routable IP address through DHCP secondary management CID, ZigBee coordinator is addressable through the VID and PID on the connected USB port, and sensors are accessible through internal 16-bit short address assigned by the coordinator when they first join the network. Therefore, for each sensor, its addressing convention is

```
short-address@zigbee/VendorID-ProductID@USB/<SS's IP address>.
```

Assuming that `9.186.1.133` is the IP address of WiMax SS, Table 4.7 summarizes the result.

4.5.2 A Device Registration Portal

We implement a Web-based registration portal to support service registration functionality, assuming that the discovery phase has been achieved by an individual platform already like ETSI M2M or OPC-UA. In our case, WiMax uses CIDs for service discovery. It is worth noting that the provided RESTful interfaces fully support the automatic registration procedure from these legacy systems, and for now we show a way for manual configurations. As shown in Figure 4.8, authorized users can log-in the system and create, view, or remove device types. Then, the user creates device domain information as in Figure 4.9(a), and registers new devices as in Figure 4.9(b). Since the device domain is already configured, when adding a new device, the user simply selects a domain node the device belongs to, either geographically or logically, as shown in Figure 4.10(a). Finally, device group is managed by adding registered devices to a group; see Figure 4.10(b). All these operations are achieved by calling the provided RESTful APIs, as shown in Section 4.4.1.

As an example, Figure 4.11 shows the XML body attached with the RESTful request `POST: http://<serverURL>/devices` to create the device profile

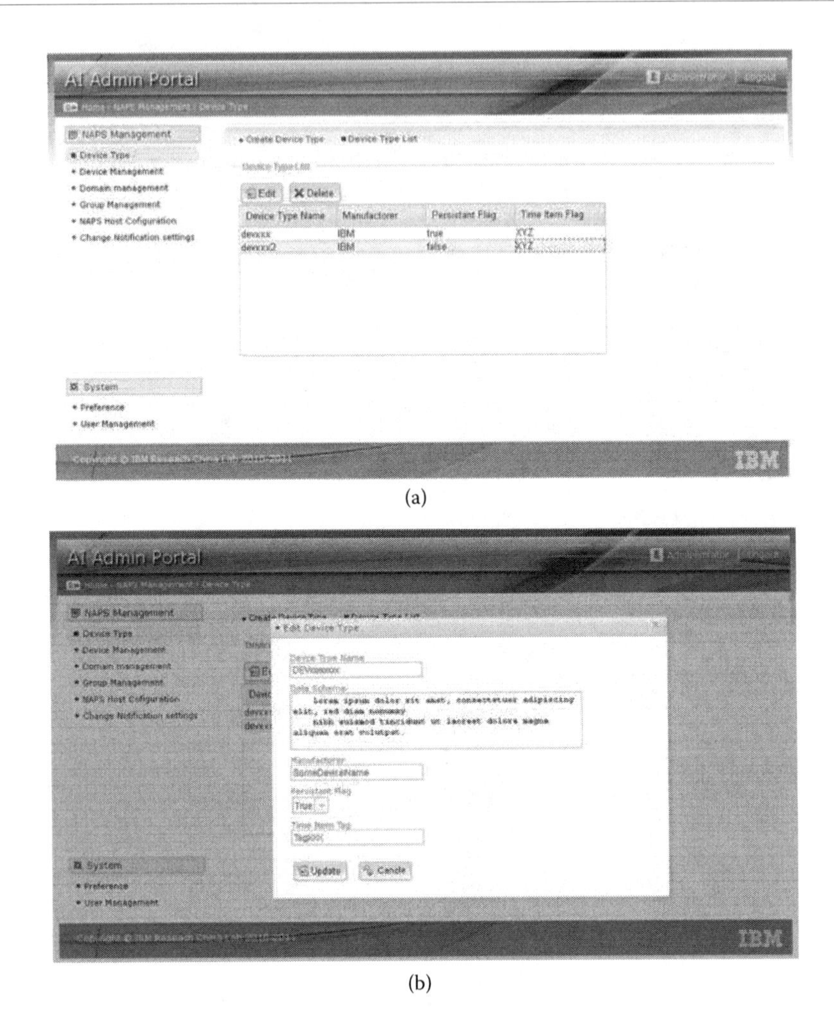

Figure 4.8: NAPS registration portal. (a) View a list of device types, and (b) add a new device type.

for temperature sensor 1. Device name, device address, location, domain information are all encoded.

4.6 Performance Evaluation

Since the previous section has verified the correctness of our model and system by deploying six sensors as a case study, in this section, we aim to evaluate

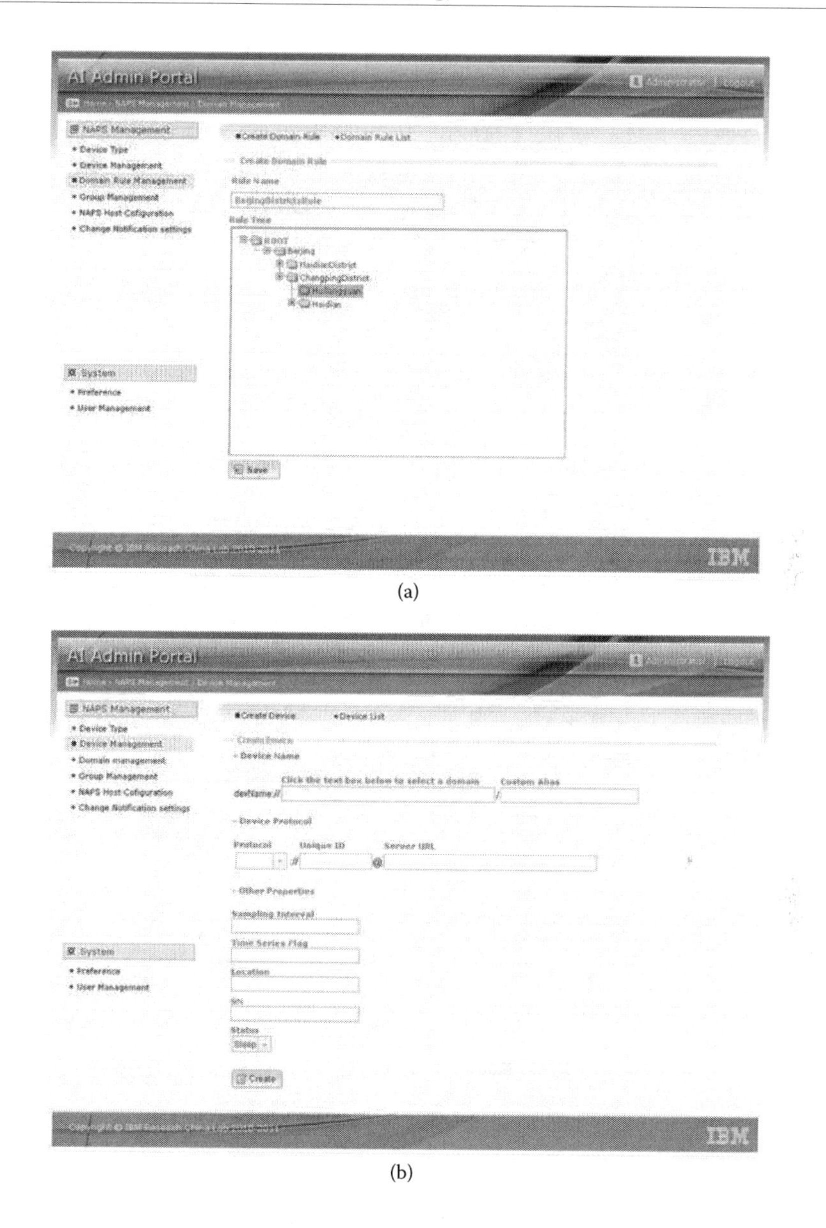

Figure 4.9: NAPS registration portal. (a) Add/view a new device domain in a tree structure, and (b) register a new device profile.

the server performance in terms of throughput when the number of devices reaches a certain amount. We testify two offered external interfaces, namely: (a) JDBC API for large amount of device meta-data retrieval (more in the configuration phase), and (b) RESTful API for information query at runtime.

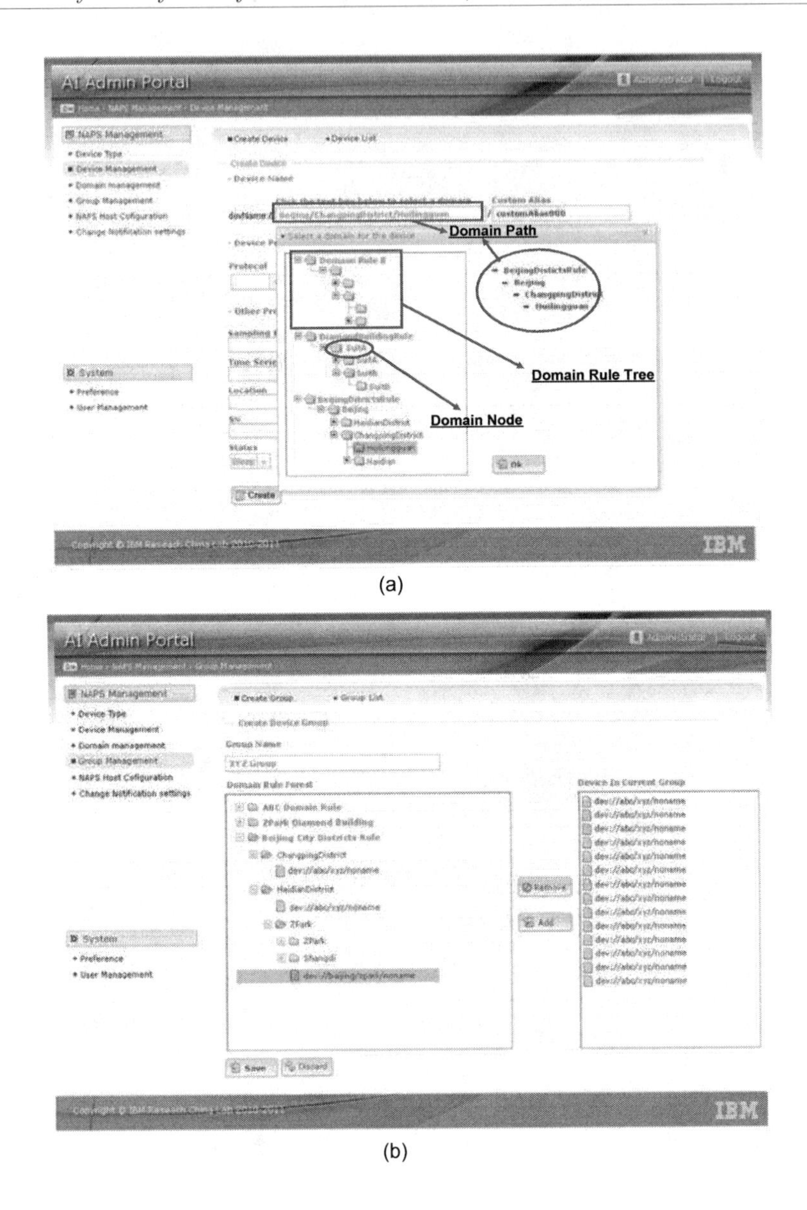

Figure 4.10: NAPS registration portal. (a) Select the device domain information from the existing domain tree, and (b) add/remove devices to/from a device group.

To run the JDBC API pressure test, we set up the environment by installing the IBM DB2, client side code (in Java Eclipse Galileo 3.5.0) on the same Thinkpad W500 workstation; and therefore the computational costs come more from the I/O side instead of the network. Furthermore, we set the JVM

```
<?xml version="1.0" encoding="UTF-8"?>
<tns:req xmlns:tns="http://IoTNAPS.org/"
      xmlns:xsi="http://www.w3.org/2001/XMLSchema-instance"
      xsi:schemaLocation="http://IoTNAPS.org/req_createDevProfile.xsd ">
 <tns:number>1</tns:number>
 <tns:profile>
  <tns:devName>dev://rm308.level3.IBM/temperature/DER452SA</tns:devName>
  <tns:address>0x4FE2@zigbee/03F0-2D12@USB/9.186.1.133</tns:address>
  <tns:protocol>zigbee</tns:protocol>
  <tns:devTypeID>21</tns:devTypeID>        <!-- temperature: (010101)B
  <tns:samplingInterval>4</tns:samplingInterval>
  <tns:timeSeriesFlag>true</tns:timeSeriesFlag>
  <tns:location>rm308.floor3.IBM</tns:location>
  <tns:sn>DER452SA</tns:sn>
  <tns:domainRuleID>5</tns:domainRuleID>  <!-- domain rule ``IBM site``: (00101)B
  <tns:domain>rm308.floor3.IBM</tns:domain>
  <tns:status>1</tns:status>
  <tns:devID_old>DER452SA</tns:devID_old>
 </tns:profile>
</tns:req>
```


Figure 4.11: The XML body attached with the RESTful URI to register a new device profile for temperature sensor 1.

heap size to 512MB for Java object creation in batch. When running the tests, we call the JDBC API to fetch the device profile records in different batch sizes, where each record is approximately 20KB. Figure 4.12 shows experimental results. For fixed batch size (e.g., 10,000 records), we observe a relatively linear increase of system response time while increasing the total number of device records in the database, showing our system scales well within the number of devices in an CPS environment. For fixed total number of devices, the response time also increases steadily with different batch sizes. Meanwhile, if one aims to retrieve five million device profiles (approximately 100GB in total), the average response time is around 33 minutes, equivalent to the system throughput around 404Mbps.

We next show the performance of RESTful APIs, and Table 4.8 demonstrates the pressure test result when searching devices under conditions. We use Apache JMeter [95], and implement the client side code (in Java Eclipse Galileo 3.5.0) on a Lenovo Thinkpad W500 workstation. The station connects to the internal WLAN (interface card supports IEEE 802.11b/g). Server is implemented on an IBM ThinkCentre M58p machine, with 1Gbps Ethernet connection, so that the client and server are within the same LAN, and wireless network should cost most. The used URI is `GET http://9.186.x.x:8080/search?devname={pollutant}`, i.e., to search for all pollutant sensors. We observe that after the system experiences some fluctuations it quickly reaches the steady state, the average achievable throughput is around 500-600 transactions per second (tps), i.e., 500-600 transactions can be simultaneously and successfully processed with responses. The throughput slightly decreases

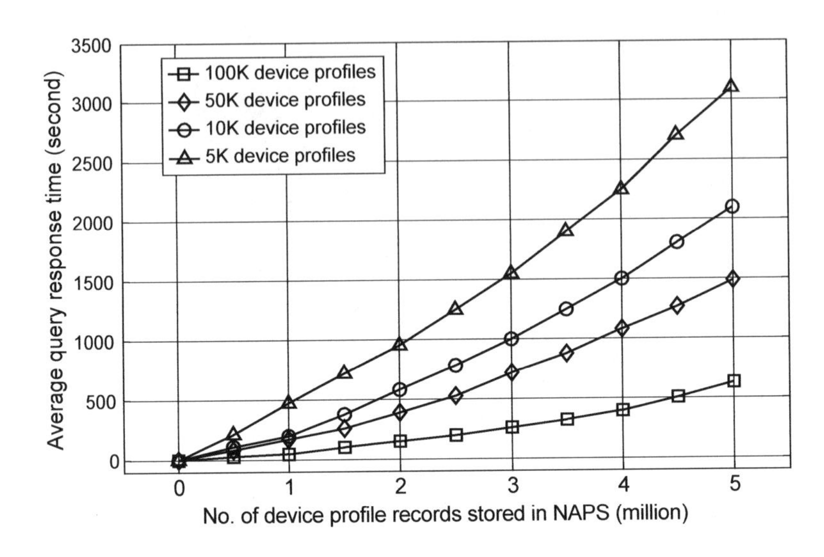

Figure 4.12: Average JDBC query response time versus the number of device profile records stored in NAPS, w.r.t. to different periodic batch fetching size.

when increasing the total number of device records. Meanwhile, the system response time increases slightly that scales well with the total number of records, and to search devices within five million records (100GB), and average response time is around 61ms.

4.7 Discussion

4.7.1 DDoS Attacks

Apart from AAA, security issues like DDoS protection should also be considered. Although leaving detailed investigations for the future, we review and discuss a few well-known mechanisms to prevent DDoS attacks. First is ingress filtering [96], where packets coming into the network are filtered if inconsistent with the source IP addresses. Second is the router throttle mechanism [97] proposed at the routers close to the victim, to proactively regulate incoming packets to a moderate level, thus reducing the amount of flooding traffic. Similarly, the key idea of pushback is to identify and control high bandwidth aggregates in network [98]. This upstream rate-limiting is called pushback and can be propagated recursively to routers further upstream. Third, since the source addresses of flooding packets are faked, various traceback techniques [99] have been proposed to find out the origin of a flooding source. Finally,

Table 4.8: Pressure Test Results on RESTful API to Search Device Names Containing Keyword "ibm"

Device record	No. of requests	avr (ms)	median (ms)	90% line (ms)	min (ms)	max (ms)	thrpt (tps)
10k	264644	17	11	29	2	3042	584.60
500k	287615	38	27	46	1	7069	544.51
1M	277728	48	36	54	2	8183	535.81
5M	255494	61	52	72	1	22551	508.26

based on the distinct protocol behavior of TCP connection establishment and teardown, the TCP SYN flooding is detected [100].

Meanwhile, commercial solutions are also available particularly for M2M scenarios such as the one provided by du Telecom [101] through a layered network. The scalability of their protect services can detect and handle threats of flexible size, using multiple methods of detection and mitigation.

4.7.2 Compatibility with IPv6

As discussed earlier in the scope of this work, we are not devising a new transport layer protocol, but aim to inter-operate with heterogeneous CPS platforms. In this sense, IPv6 based networks (like 6LowPan) is one kind of such solutions to be integrated with our NAPS middleware. As presented in this chapter, NAPS performs the uniform name-to-address resolution, comparable to the functionality of DNS in the IP world. Furthermore, it provides abundant RESTful APIs for device, device type, and device group profile services. During the downstream command delivery, once the addressing is obtained from NAPS, it will forward it to the corresponding Internet gateway, such as the M2M gateway in ETSI M2M service architecture, OPC-UA server, or WiMax base station (as considered in the case study). In a hierarchy, these gateways further route the command to the next level gateway (e.g., the ZigBee coordinator) which maintains its own addressing mechanism to the device. In this way, we do not attempt to "escalate" the IPv6 naming up to the middleware layer, and are thus able to go around the issue of potential insufficiency problems when the number of devices is massive beyond the capacity of IPv6.

4.8 Summary

In this chapter, we presented NAPS, a middleware to support device naming, application addressing and profile storage and look-up services in CPS sensory environments. Different from all previous efforts only focusing on a specific standard/protocol, our design can work with any existing system and platform to assist the upstream data collection and identification, content-based data filtering and matching, downstream control message delivery, as well as application query. Our contributions were the proposal of such a complete and detailed design, including its key functionalities, system flows, interfaces, and individual module designs, the proposal of a unique device naming and addressing convention with applicability to widely used standards and protocols, the proposal of an efficient identifier generation scheme, and finally a real case study in a smart building environment. Performance evaluation showed that a single node commodity server can achieve average throughput of around 500-600tps, with system response time 61ms to search devices within five million devices.

Chapter 5

Device Search and Selection for CPS

Charith Perera

Open University, U.K.

Chi Harold Liu

Beijing Institute of Technology, China

Peter Christen

Australian National University, Australia

CONTENTS

Besides mobile sensing platforms and naming/addressing/profile services, searching functionality also plays a vital role in the domain of cyber-physical systems (CPS). Many different types of search capabilities are required to build a comprehensive CPS architecture. In CPS, users may want to search smart devices and services. In this chapter, we discuss concepts and techniques related to device search and selection. We briefly discuss different types of device searching approaches where each has its own objectives and applications. One such device searching technique is context-aware searching. In this chapter, we present context-aware sensor search, selection, and ranking model called CASSARAM in detail. This model addresses the challenge of efficiently selecting a subset of relevant sensors out of a large set of sensors with similar functionality and capabilities. CASSARAM takes into account user preferences and considers a broad range of sensor characteristics, such as reliability, accuracy, location, battery life, and many more. Later in the chapter, we discuss three different techniques that can be used to improve the efficiently of CASSARAM. We implemented the proof of concept software using Java. Testing and performance evaluation results are also discussed. We also highlight open research challenges and opportunities in order to support future research directions.

5.1 Introduction

When a large number of sensors are available from which to choose, it becomes a challenge and a time-consuming task to select the appropriate sensors that will help the users to solve their own problems. Besides, when more and more sensors get connected to the Internet, the search functionality becomes critical. This chapter addresses the problem mentioned above as we observe the lack of focus on sensor selection and search in existing CPS solutions and research. Some of the early work on searching in sensor networks is presented in [102].

In this chapter, we discuss a model that provides context-aware sensor search support to any CPS middleware solution. Our contributions can be summarized as follows. We have developed an ontology-based context framework for sensing devices in CPS which allows capturing and modelling context properties related to sensors. This information allows users to search the sensors based on context. We have designed, implemented, and evaluated the proposed CASSARAM model and its performance in a comprehensive manner. Specifically, we propose a comparative-priority based weighted index (CPWI) technique to index and rank sensors based on the user preferences. Furthermore, we propose a comparative-priority based heuristic filtering (CPHF) technique to make the sensor search process more efficient. We also propose a relational-expression based filtering (REF) technique to support more comprehensive searching. Finally, we propose and compare several distributed sensor search mechanisms. During this chapter, we mainly focused on searching the sensors. However, as we mentioned above, sensors are always attached to sensing devices. Therefore, once the required sensors are identified through search, the actual communication that is required to retrieve data needs to be done with the respective sensing devices.

Hereafter, the chapter is organized in the following way: Section 5.2 briefly explains the architecture of CPS and the importance of searching functionality toward the architecture. Further, it discusses variety of different ways of searching sensing devices in CPS paradigm. It also explains how each searching technique uniquely contributes to a comprehensive sensing strategy. Then, we introduce the problem of context-aware device search in Section 5.3. In Section 5.4, we propose our solution, context-aware sensor search and selection model (CASSARAM). Then, we introduce three techniques that can be used to enhance and enrich the performance of CASSARAM in Section 5.5. Implementation and evaluation details are presented in Sections 5.6 and 5.7, respectively. Finally, we highlight open research challenges and opportunities in this domain in Section 5.8. The concluding remarks are presented in Section 5.9.

5.2 Internet of Things Architecture and Search Functionality

In this section, we briefly introduce the Internet of Things architecture and the applicability of device search functionality toward CPS middleware. However, we are not going to discuss a concrete architecture. Instead, we highlight the relevance and importance of different types of search functionalities. *"Middleware is a software layer that stands between the networked operating system and the application and provides well known reusable solutions to frequently encountered problems like heterogeneity, interoperability, security, dependability"* [103]. The functionalities required by CPS middleware platforms are explained in detail in [104, 105, 106, 107, 28]. Further, challenges in developing middleware solutions for the CPS are discussed in [108]. In a CPS middleware platform, searching functionality may mainly be used to search CPS resources (smart devices, services, people) [109].

In this chapter, we concentrate on searching smart devices. Such search functionality is used by users as well as applications to support both machine-to-machine communications and sensing as service models. Some of the major challenges such as the dynamic nature, massive number of devices, real-time searching, and privacy in searching are discussed in [110]. Further, radio-frequency identification (RFID) enabled objects make the searching hard. Especially, unlike the Web which is built of static documents, sensor reading possesses a very short life span and real-world entities are highly dynamic. This implies the traditional search techniques of the Web do not work in CPS.

5.2.1 Sensing Device Searching from Functional Perspective

Smart sensing devices can be searched using different techniques. There are three primary techniques to search sensing devices: thematic, spatial, and temporal [111]. Additionally, there are a few other ways to search sensing devices as discussed below. It is important to note that these techniques can be combined to provide comprehensive CPS middleware solution.

- **Thematic searching** is finding the sensing devices based on the phenomena. For example, a user may want to find devices that sense temperature or pressure. This category of techniques may extend to search sensing devices with capabilities to sense multiple phenomena. This is one of the most the widely used searching techniques.

- **Spatial searching** is finding sensors that are in a particular geographical location or area. For example, a user may want to find devices that are in Canberra. The location can be based on geographical coordinates. It can also be based on high-level abstractions such as *Australian National University.* Mayer et al. [112] consider the location of sensing devices as the main context property and structure them in a logical tree structure. The sensing devices are by location using a tree search technique. Search queries are distributively processed in different paths/nodes of the tree. Mayer et al. have represented the layers of abstractions using a tree structure. For example, a root node represents the whole building. Children nodes of the root can thus represent different floors, and rooms of each floor can be denoted by children's children.

Shakkottai et al. [113] address the problem of a user searching for information over a sensor network, where the user does not have prior knowledge of the location of the information. They have used three strategies to search sensors. First, query (i.e., packet) moves from one node to another until it finds a match. In the second strategy, sensors advertise their information and queries also move through nodes looking for a match. In the final method, sensors share the information (i.e., their capabilities) with each other which makes it easier for the query to find a match as the distributed information can guide the query to a matching sensor much faster.

- **Temporal searching** is finding sensors that offer data for a set point in time or period. For example, a user may need to find sensing devices that deliver data measured during the last month. Another example would be when a user searches sensing devices that are capable of providing sensor data for the next week at a sampling rate of 30 seconds.

- **Content-based searching** is finding the sensors that produce (or have produced in the past) a certain pattern of values. For example, a user may want to monitor environmental pollution. So he may search sensors that produce certain patterns of sensor data values. The user may collect data from those sensors for his own application (e.g., for further processing in the future). Truong et al. [114] propose a fuzzy-based technique to search sensing devices. The technique generates similarity scores to compare the output of a given sensor with the outputs of several other sensors in order to find a similar sensor. Elahi et al. [115] propose a content-based sensor search approach (i.e., finding a sensor that outputs a given value at the time of a query).

  Dyser is a search engine proposed by Ostermaier et al. [116] for real-time Internet of Things, which uses statistical models to make predictions about the state of its registered objects (sensors). When a user submits a query, *Dyser* pulls the latest data to identify the actual current state to decide whether it matches the user query. Prediction models help to find matching sensors with a minimum number of sensor data retrievals. This work also searches sensing devices based on content, but it uses predictive models to predict content (e.g., predict the state of a given sensor at a time based on historic data) instead of using real content in order to support faster real-time searching. A survey on real time searching in CPS paradigm is presented in [117]. They have compared eight different CPS middleware solutions in terms of their searching functionality.

- **Cluster searching** is a combination of spatial and thematic searching techniques that find clusters of devices. For example, a user may need to find devices that sense temperature, air pressure, and luminosity where all the sensing devices are located within five meters' radius from each other. Such close proximity allows algorithms to fuse, and reason data assuming that all devices are approximately deployed together.

- **Context-aware searching** is finding the sensing devices based on context information related to those devices. Some of the context information that can be used to search sensing devices are availability, accuracy, latency, battery life, etc. In this chapter, we focus on searching devices based on context. A comprehensive set of context information that can be used to search sensing devices is discussed later in the Section 5.4.6. For example, this type of searching may be useful to find devices that have more energy remaining when more than one device that can perform the required sensing task is available. Algorithms can put the device with less energy into sleep mode and retrieve data from the device which has more energy. Such optimization will increase the life-time of the entire sensing device network.

- **Manual interactive searching** is allowing users to select sensing devices manually from which they want to retrieve data through different kinds of user interfaces. Users may be provided with graphical user interface (e.g., may be Web based or may be interactive touch-based walls, tables or tabletops such a Pixelsense[1]) where devices are mapped based on location (either outdoor or indoor maps). Users may pick sensors by selecting sensing devices one by one. Additionally, users may mark certain geographical areas, so the sensors within the marked area get selected. Such techniques will complement other techniques, so the users have more control and it allows fine tuning the search results. Noguchi et al. [118] has proposed a framework that visualizes sensors and sensor data on a map through a graphical user interface.

5.2.2 Sensing Device Searching from Implementation Perspective

In addition to the above-mentioned searching techniques that we discussed from a functional perspective, implementation of each technique can also be varied. Additionally, there can be differences in data storage models (i.e., text, metadata, semantic), time taken to process (i.e., real-time or archival), location of the data storage, and so on. Let us discuss some of the widely used techniques.

- **Text-based search:** One of the primary ways of searching sensing devices is to describe the sensor's capabilities and functionalities in clear text and use a search engine to search relevant sensors. By extending this concept, Snoogle [119], a search engine for pervasive computing

[1] www.microsoft.com/en-us/pixelsense/

environments built on a network of smart devices, stores a textual description of that object in the form of keywords. Through a layered architecture, Snoogle allows users to describe their requirements using keywords. Snoogle searches appropriate object sensors matching the keywords. However, such traditional Web search-like approach does not work accurately in the CPS sensor selection and search domain as text based search approaches cannot capture the critical characteristics of a sensor accurately.

- **Metadata-based search:** Another approach that can be followed is metadata annotation. Even if we maintain metadata on the sensors (e.g., stored in a sensor's storage) or in the cloud, interoperability will be a significant issue. Furthermore, a user study done by Broring et al. [120] has described how 20 participants were asked to enter metadata for a weather station sensor using a simple user interface. Those 20 people made 45 mistakes in total. The requirement of re-entering metadata in different places (e.g., entering metadata on GSN [58] once and again entering metadata on OpenIoT [121], etc.) arises when we do not have common descriptions.

- **Ontology-based search:** Recently, the W3C Incubator Group released Semantic Sensor Network XG Final Report, which defines an SSN ontology [122]. The SSN ontology allows describing sensors, including their characteristics. This effort increases the interoperability and accuracy due to the lack of manual data entering. Furthermore, such mistakes can be avoided by letting the sensor hardware manufacturer produce and make available sensor descriptions using ontologies so that CPS solution developers can retrieve and incorporate (e.g., mapping) them in their own software system. Paparrizos et al. [123] provides an easy-to-use query interface, built upon semantic technologies where users can freely store and query their metadata. Location information sensor types can be provided though form-based Web user interface.

  Based on the arguments above, ontology-based sensor description and data modelling is useful for CPS solutions. This approach also allows semantic querying. Our proposed solution allows the users to express their priorities in terms of sensor characteristics and it will search and select appropriate sensors. In our model, both quantitative reasoning and semantic querying techniques are employed to increase the performance of the system by utilizing the strengths of both techniques.

- **Distributed search:** Sensing devices in CPS paradigm may be connected to different middleware solutions and server nodes. Therefore, distributed processing is required to find sensing devices from multiple middleware instances. Microsearch [124] is a search system suitable for small devices (e.g., a 12MHz CPU, 64KB of RAM, 512KB of ash memory, and wireless capabilities, all packaged in a 3x3 cm circuit board) used in ubiquitous computing environments. Microsearch allows us to index static information stored in sensor nodes in distributed manner and to search for sensor nodes that store a particular static information.

- **Real-time search:** As we discussed earlier, *Dyser* [116] uses predictive models to predict content (e.g., predict the state of a given sensor at a time based on historic data) instead of using real content in order to support faster real-time searching.

There are a number of existing solutions that have been proposed and developed in this area. Let us discuss some of these leading and interesting solutions that will help to understand our solution presented in later sections. Ideally, CPS middleware solutions should allow the users to express what they want and provide the relevant sensor data back to them quickly without asking the users to manually select the sensors which are relevant to their requirements. Even though CPS has received significant attention from both academia and industry, sensor search and selection has not been addressed comprehensively. Specifically, sensor search and selection techniques using context information [125] have not been explored substantially. A survey on context aware computing for the Internet of Things [125] has recognized sensor search and selection as a critical task in automated sensor configuration and context discovery processes. Another review on semantics for the Internet of Things [126] has also recognized resource (e.g., a sensor or an actuator) search and discovery functionality as one of the most important functionalities required in CPS. Barnaghi et al. [126] have highlighted the need for semantic annotation of CPS resources and services. Processing and analyzing the semantically annotated data are essential elements to support search and discovery [126]. This justifies our approach of annotating the sensors with related context information and using that to search the sensors. The following examples show how existing CPS middleware solutions provide sensor searching functionality.

Linked Sensor Middleware (*LSM*) [127, 128] provides some sensor selection and searching functionality. However, they have very limited capabilities, such as selecting sensors based on location and sensor type. All the searching needs to be done using SPARQL, which is not user-friendly to non-technical users. Similar to *LSM*, there are several other CPS middleware related projects under development at the moment. GSN [58] is a platform aiming at providing flexible middleware to address the challenges of sensor data integration and distributed query processing. It is a generic data stream processing engine. GSN has gone beyond the traditional sensor network research efforts such as routing, data aggregation, and energy optimization. GSN lists all the available sensors in a combo-box which users need to select. However, GSN lacks semantics to model the metadata. Another approach is Microsoft *Sen-*

sorMap [129]. It only allows users to select sensors by using a location map, by sensor type, and by keywords. *xively* (xively.com) is also another approach which provides a secure, scalable platform that connects devices and products with applications to provide real-time control and data storage. This also provides only keyword search. The illustrations of the search functionalities provided by the above-mentioned CPS solutions are presented in [130]. Our proposed solution CASSARAM can be used to enrich all the above-mentioned CPS middleware solutions with a comprehensive sensor search and selection functionality.

According to a study in Europe [131], there are over 12,000 working and useful Web services on the Web. Even in such conditions, choice between alternatives (depending on context properties) has become a challenging problem. The similarities strengthen the argument that sensor selection is an important challenge at the same level of complexity as Web services. On the other hand, the differences show that sensor selection will become a much more complex challenge over the coming decade due to the scale of the CPS. In the following, we briefly describe some of the work done in sensor searching and selection.

De et al. [132] have proposed a conceptual architecture, a CPS platform, to support real-world and digital objects. They have presented several semantic ontology-based models that allow capturing information related to CPS resources (e.g., sensors, services, actuators). However, they are not focused on sensors and the only context information considered is location. In contrast, CASSARAM narrowly focuses on sensors and considers a comprehensive set of context information (see Section 5.4.6).

Guinard et al. [133] have proposed a Web service discovery, query, selection, and ranking approach using context information related to the CPS domain. Similarly, TRENDY [134] is a registry-based service discovery protocol based on CoAP (Constrained Application Protocol) [135] based Web services with context awareness. This protocol has been proposed to be used in the Web of Things (WoT) domain with the objective of dealing with a massive number of Web services (e.g., sensors wrapped in Web services). Context information such as hit count, battery, and response time are used to select the services. An interesting proposal is by Calbimonte et al. [136], who have proposed an ontology-based approach for providing data access and query capabilities to streaming data sources. This work allows the users to express their needs at a conceptual level, independent of implementation. Our approach, CASSARAM, can be used to complement their work where we support context-based sensor search and they provide access to semantically enriched sensor data. Furthermore, our evaluation results can be used to understand the scalability and computational performance of their working big data paradigm as both approaches use the SSN ontology.

Garcia-Castro et al. [137] have defined a core ontological model for semantic sensor Web infrastructures. It can be used to model sensor networks (by extending the SSN ontology), sensor data sources, and the Web services that expose the data sources. Our approach can also be integrated into the *uBox*

[138] approach, to search *things* in the WoT domain using context information. Currently, *uBox* performs searches based on location tags and object (sensor) classes (types) (e.g., hierarchy local/class/actuator/light). As we mentioned earlier, collecting data from all the available sensor devices is not required all the time. Further, such methods are inefficient in terms of resource consumption both in sensor network domain and CPS paradigm. The cost models that can be used to measure the cost of search are discussed in [139]. IoT-SVK [140] is a hybrid search engine framework for the CPS paradigm, and it supports three search modes: keyword-based searches, spatial-temporal searches, and value-based searches.

5.3 Problem Definition

The problem that we address in this chapter can be defined as follows. Due to the increasing number of sensors available, we need to search and select sensors that provide data which will help to solve the problem at hand in the most efficient and effective way. Our objective is not to solve the users' problems, but to help them to collect sensor data. The users can further process such data in their own ways to solve their problems. In order to achieve this, we need to search and select sensors based on different pieces of context information. Mainly, we identify two categories of requirements: point-based requirements (non-negotiable) and proximity-based (negotiable) requirements. We examined the problem in detail in [130] by providing real-world application scenarios and challenges.

First, there are the point-based requirements that need to be definitely fulfilled. For example, if a user is interested in measuring the temperature in a certain location (e.g., Canberra), the result (e.g., the list of sensors) should only contain sensors that can measure temperature. The user cannot be satisfied by being providing with any other type of sensor (e.g., pressure sensors). There is no bargain or compromise in this type of requirement. Location can be identified as a point-based requirement. The second is proximity-based requirements that need be fulfilled in the best possible way. However, meeting the exact user requirement is not required. Users may be willing to be satisfied with a slight difference (variation). For example, the user has the same interest as before. However, in this situation, the user imposes proximity-based requirements in addition to point-based requirements. The user may request sensors having an accuracy of around 92%, and reliability 85%. Therefore, the user gives the highest priority to these characteristics. The user will accept sensors that closely fulfill these requirements even though all other characteristics may not be favorable (e.g., the cost of acquisition may be high and the sensor response may be slow). It is important to note that users may not be able to provide any specific value, so the system should be able to understand the user's priorities and provide the results accordingly, by using comparison techniques.

Another motivation behind our research are statistics and predictions that show rapid growth in sensor deployment related to the CPS and smart cities. It is estimated that today there about 1.5 billion Internet-enabled PCs and over 1 billion Internet-enabled mobile phones. By 2020, there will be 50 to 100 billion devices connected to the Internet [28]. Furthermore, our work is motivated by the increasing trend of CPS middleware solutions development. Today, most of the leading middleware solutions provide only limited sensor search and selection functionality.

We highlight the importance of sensor search functionality using current and potential applications. Smart agriculture [141] projects such as Phenonet [142] collect data from thousands of sensors. Due to heterogeneity, each sensor may have different context values, as mentioned in Section 5.4.6. Context information can be used to selectively select sensors depending on the requirements and situations. For example, CASSARAM helps to retrieve data only from sensors which have more energy remaining when alternative sensors are available. Such action helps to run the entire sensor network for a much longer time without reconfiguring and recharging. The sensing as a service [33] architectural model envisions an era where sensor data will be published and sold through the cloud. Consumers (i.e., users) will be allowed to select a number of sensors and retrieve data for some period as specified in an agreement by paying a fee. In such circumstances, allowing consumers to select the sensors they want based on context information is critical. For example, some consumers may be willing to pay more for highly accurate data (i.e., highly accurate sensors) while others may be willing to pay less for less accurate data, depending on their requirements, situations, and preferences.

5.4 Context-Aware Approach for Device Search and Selection

In this section, we present the proposed sensor selection approach step by step in detail. First, we provide a high-level overview of the model, which describes the overall execution flow and critical steps. Then, we explain how user preferences are captured. Next, the data representation model and proposed extensions are presented. Finally, the techniques of semantic querying and quantitative reasoning are discussed with the help of some algorithms. All the algorithms presented in this chapter are self-explanatory and the common algorithmic notations used in this chapter are presented in Table 5.1.

5.4.1 *High-Level Model Overview*

The critical steps of CASSARAM are presented in Figure 5.1. As we mentioned earlier our objective is to allow the users to search and select the sensors that best suit their requirements. In our model, we divide user

Table 5.1: Common Algorithmic Notation Table

Symbol	Definition
$\mathbb{O}$	*Ontology* consists of sensor descriptions and context property values related to all sensors
$\mathbb{P}$	*UserPrioritySet* contains user priority value for all context properties
$\mathbb{Q}$	*Query* consists of point-based requirements expressed in SPARQL
N/N_{All}	Number of sensors required by the user / Total number of sensors available
$\mathbb{S}_{Filtered}$	This contains the results of the query $\mathbb{Q}$
$\mathbb{S}_{Results}$	*ResultsSet* contains selected number of sensors
$\mathbb{S}_{Indexed}$	*IndexedSensorSet* stores the index values of the sensors
$\mathbb{M}$	Multidimensional space where each context property is represented by a dimension and sensors are plotted
$\mathbb{UI}$	*UserInput* consists of input values provided by the users via the user interface
$\mathbb{SC}/SC$	Values of all the sliders / Value of a slider
$\mathbb{P}^w$	This contains user priority value converted into weights using normalization
p_i/p_i^w	Value of ith context property / Value of ith context property in normalized form
$\mathbb{CP}/CP$	*ContextPropertySet* consists of all context information / value of ith context property
$\mathbb{NCP}$	Normalized Context Property Set
M	Margin of error
S_j	This is the jth sensor
$CP_i^{S_j}$	CP value of ith property of jth sensor.
CP^{ideal}	CP values of the ideal sensors that user prefers

requirements into two categories (from the user's perspective): *point-based requirements* and *proximity-based requirements*, as discussed in Section 5.3. Algorithm 1 describes the execution flow of CASSARAM. At the beginning,

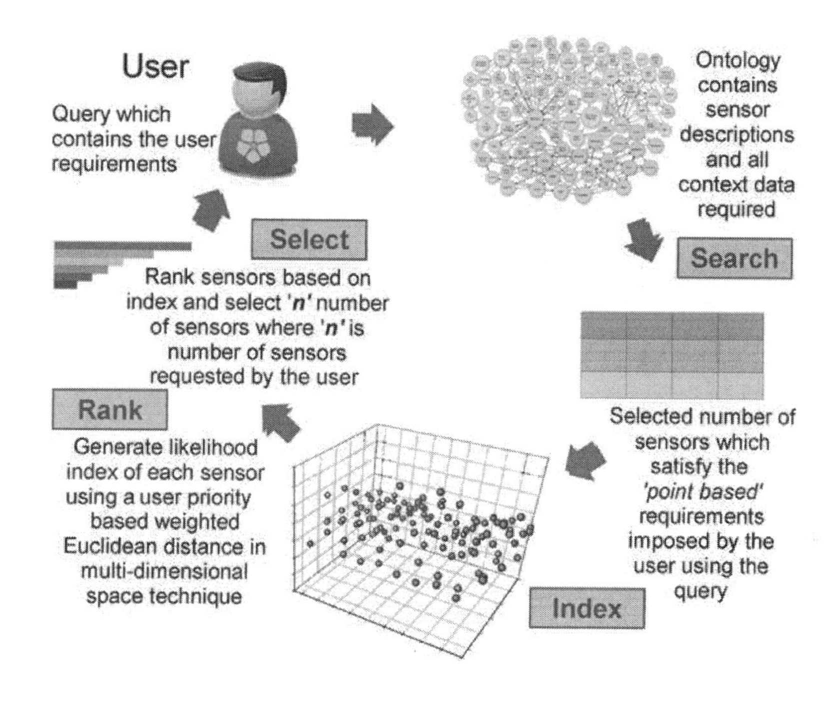

Figure 5.1: High level overview of CASSARAM.

CASSARAM identifies the point-based requirements, the proximity-based requirements, and the user priorities. First, users need to select the point-based requirements. For example, a user may want to collect sensor data from 1,000 temperature sensors deployed in Canberra. In this situation, the sensor type (i.e., temperature), location (i.e., Canberra) and number of sensors required (i.e., 1,000) are the point-based requirements. Our CASSARAM prototype tool provides a user interface to express this information via SPARQL queries. In CASSARAM, any context property can become a point-based requirement. Next, users can define the proximity-based requirements. All the context properties we will present in Section 5.4.6 are available to be defined in comparative fashion by setting the priorities via a slider-based user interface, as depicted in Figure 5.2. Next, each sensor is plotted in a multi-dimensional space where each dimension represents a context property (e.g., accuracy, reliability, latency). Each dimension is normalized [0,1] as explained in Algorithm 3. Then, the Comparative-Priority Based Weighted Index (CPWI) is generated for each sensor by combining the user's priorities and context property values as explained in Section 5.4.5. The sensors are ranked using the CPWI and the number of sensors required by the user is selected from the top of the list.

Algorithm 1 Execution Flow of CASSARAM

Require: $(\mathbb{O})$, $(\mathbb{P})$, $(\mathbb{Q})$, (N), $(\mathbb{M})$.
1: **Output:** $\mathbb{S}_{Results}$
2: $\mathbb{S}_{Filtered} \leftarrow$ queryOntology$(\mathbb{O}, \mathbb{Q})$
3: **if** cardinality$(\mathbb{S}_{Filtered}) < N$ **then**
4: **return** $\mathbb{S}_{Results} \leftarrow \mathbb{S}_{Filtered}$
5: **else**
6: $\mathbb{P} \leftarrow$ capture user priorities$(\mathbb{UI})$
7: $\mathbb{M} \leftarrow$ Plot sensors in multidimensional space$(\mathbb{S}_{Results})$
8: $\mathbb{S}_{Indexed} \leftarrow$ calculate CPWI$(\mathbb{S}_{Results}, \mathbb{M})$
9: $\mathbb{S}_{Results} \leftarrow$ rank sensors$(\mathbb{S}_{Indexed})$
10: $\mathbb{S}_{Results} \leftarrow$ select sensors$(\mathbb{S}_{Results}, N)$
11: **return** $\mathbb{S}_{Results}$
12: **end if**

5.4.2 Capturing User Priorities

This is a technique we developed to capture the user's priorities through a user interface, as shown in Figure 5.2. CASSARAM allows users to express which context property is more important to them, when compared to others. If a user does not want a specific context property to be considered in the indexing process, they can avoid it by not selecting the check-box correlated with that specific context property. For example, according to Figure 5.2, *energy* will not be considered when calculating the CPWI. This means the user is willing to accept sensors with any *energy* consumption level. Users need to position the slider of each context property if that context property is important to them. The slider scale begins from 1, which means no priority (i.e., the left corner). The highest priority can be set by the user as necessary with the help of a *scaler*, where a higher scale makes the sliders more sensitive (e.g., $10^2 =$ 1 to $100, 10^3, 10^4$). Algorithm 2 describes the user priority capturing process.

As depicted in Figure 5.2, if the user wants more weight to be placed on the reliability of a sensor than on its accuracy, the reliability slider needs to be placed further to the right than the accuracy slider. A weight is calculated for each context property. Therefore, higher priority means higher weight. Sensors with high reliability and accuracy will be ranked highly. However, those sensors may have high costs due to the low priority placed on cost.

5.4.3 Data Modelling and Representation

In this chapter, we employed the Semantic Sensor Network Ontology (SSN) [122] to model the sensor descriptions and context properties. The main reasons for selecting the SSN ontology are its interoperability and the trend toward ontology usage in the CPS and sensor data management domain. A comparison of different semantic sensor ontologies is presented in [59]. The SSN

Algorithm 2 User Priority Capturing

Require: $(\mathbb{UI})$, $(\mathbb{SC})$
1: **Output:** $\mathbb{P}^w$
2: $\mathbb{P} \leftarrow$ extract user priorities$(\mathbb{UI})$
3: $SC_{Highest} \leftarrow$ get maximum priority$(\mathbb{SC})$
4: $SC_{Lowest} \leftarrow$ get minimum priority$(\mathbb{SC})$
5: $SC_{Range} \leftarrow SC_{Highest} - SC_{Lowest}$
6: **for** each context property priority $p_i \in \mathbb{P}$ **do**
7: $\quad p_i^w \leftarrow (p_i \div S_{Range})$
8: $\quad$ **if** $p_i^w \geq 0$ **then**
9: $\quad\quad$ add p_i^w to $\mathbb{P}^w$
10: $\quad$ **else**
11: $\quad\quad$ continue
12: $\quad$ **end if**
13: **end for**
14: **return** $\mathbb{P}^w$

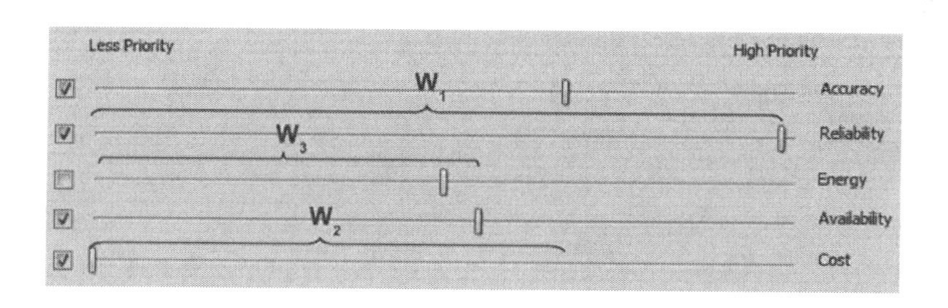

Figure 5.2: A weight of W_1 is assigned to the *reliability* property. A weight of W_2 is assigned to the *accuracy* property. A weight of W_3 is assigned to the *availability* property. A weight of W_4, the default weight, is assigned to the *cost* property. High priority means always favored, and low priority means always disfavored. For example, if the user makes *cost* a high priority (more toward the right), that means CASSARAM tries to find the sensors that produce data at the lowest cost. Similarly, if the user makes *accuracy* a high priority, that means CASSARAM tries to find the sensors that produce data with high accuracy.

ontology is capable of modelling a significant amount of information about sensors, such as sensor capabilities, performance, the conditions in which it can be used, etc. The details are presented in [122]. The SSN ontology includes the most common context properties, such as accuracy, precision, drift, sensitivity, selectivity, measurement range, detection limit, response time, frequency, and latency. However, the SSN ontology can be extended *unlimitedly* by a categorization with three classes: *measurement property*, *operating*

Algorithm 3 Flexi-Dynamic Normalization

Require: ($\mathbb{CP}$), ($\mathbb{S}$), (cp_i),
1: **Output:** $\mathbb{NCP}$
2: $cp_i^{S_j} \leftarrow$ receive new property value*
3: $cp_i^{highest} \leftarrow$ retrieve highest($\mathbb{CP}$)
4: $cp_i^{lowest} \leftarrow$ retrieve lowest($\mathbb{CP}$)
5: **if** $cp_i^{highest} < cp_i^{S_j}$ **then**
6: $\quad cp_i^{highest} \leftarrow cp_i^{S_j}$
7: $\quad$ **for** each $cp_i^{S_j} \in \mathbb{CP}, \mathbb{S}$ **do**
8: $\quad\quad$ update($\mathbb{NCP}$) $\leftarrow [\frac{(cp_i^{S_j} - cp_i^{lowest})}{(cp_i^{highest} - cp_i^{lowest})}]$
9: $\quad$ **end for**
10: **else**
11: $\quad$ update($\mathbb{NCP}$) $\leftarrow [\frac{(cp_i^{S_j} - cp_i^{lowest})}{(cp_i^{highest} - cp_i^{lowest})}]$
12: **end if**
13: **return** $\mathbb{NCP}$

*sensors registered in the CPS middleware

property, and *survival property*. We depict a simplified segment of the SSN ontology in Figure 5.3. We extend the *quality class* by adding several subclasses based on our context framework, as listed in Section 5.4.6. All context property values are stored in the SSN ontology in their original measurement units. CASSARAM normalizes them on demand to [0,1] to ensure consistency. Caching techniques can be used to increase the execution performances. Due to technological advances in sensor hardware development, it is impossible to statically define upper and lower bounds for some context properties (e.g., battery life will be improved over time due to advances in sensor hardware technologies). Therefore, we propose Algorithm 3 to dynamically normalize the context properties.

5.4.4 Filtering Using Querying Reasoning

Once the point-based requirements of the user have been identified, they need to be expressed using SPARQL. Semantic querying has weaknesses and limitations. When a query becomes complex, the performance decreases [143]. Relational expression based filtering can also be used; however, using it will increase the computational requirements. Further explanations are presented in Section 5.5.2. Any of the context properties identified in Section 5.4.6 can become point-based requirements and need to be represented in SPARQL. This step produces $\mathbb{S}_{Filtered}$, where all the sensors satisfy all the point-based requirements.

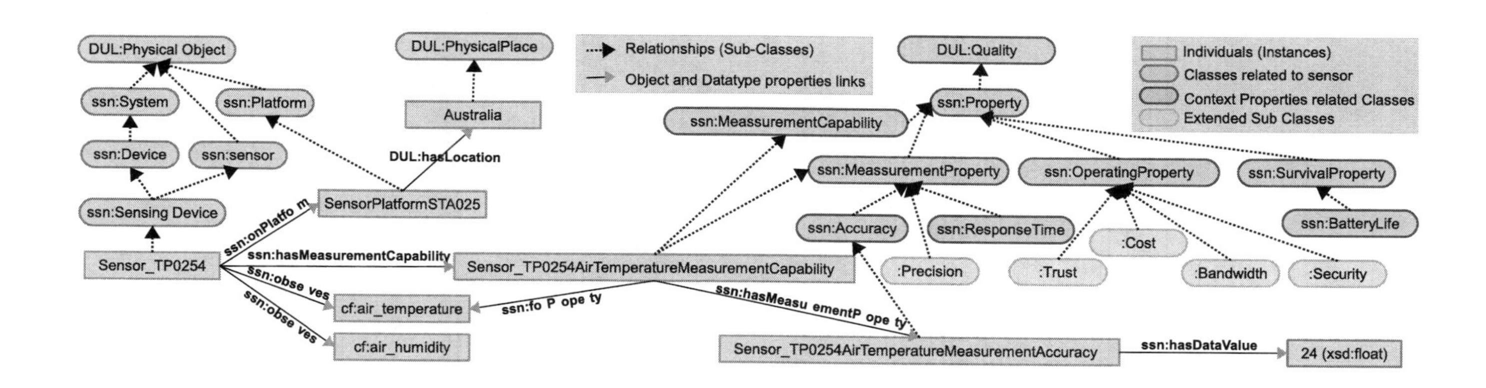

Figure 5.3: Data model used in CASSARAM. In SSN ontology, *sensors* are not constrained to physical sensing devices; rather a sensor is anything that can estimate or calculate the value of a phenomenon, so a device or computational process or combination could play the role of a sensor. A *sensing device* is a device that implements sensing [122]. *Sensing device* is also a sub-class of *sensor*. By following the above definition, our focus is on sensors. CF (Climate and Forecast) ontology is a domain specific external ontology. DOLCE+DnS Ultralite (DUL) ontology provides a set of upper level concepts that can be the basis for easier interoperability among many middle and lower level ontologies. More details are provided in [122].

5.4.5 Ranking Using Quantitative Reasoning

In this step, the sensors are ranked based on the proximity-based user requirements. We developed a weighted Euclidean distance based indexing technique, called the Comparative-Priority Based Weighted Index (CPWI), as follows.

$$(CPWI) = \sqrt{\sum_{i=1}^{n} \left[W_i(U_i^d - S_i^\alpha)^2\right]}$$

First, each sensor is plotted in multi-dimensional space where each context property is represented by a dimension. Then, users can plot an ideal sensor in the multi-dimensional space by manually entering context property values as illustrated in Figure 5.4 by U_i. By default, CASSARAM will automatically plot an ideal sensor as depicted in U_d (i.e., the highest value for all context properties). Next, the priorities defined by the user are retrieved. Based on the positions of the sliders (in Figure 5.2), weights are calculated in a comparative fashion. Algorithm 4 describes the indexing process. It calculates the CPWI and ranks the sensors using reverse-normalized techniques in descending order. CASSARAM selects N sensors from the top.

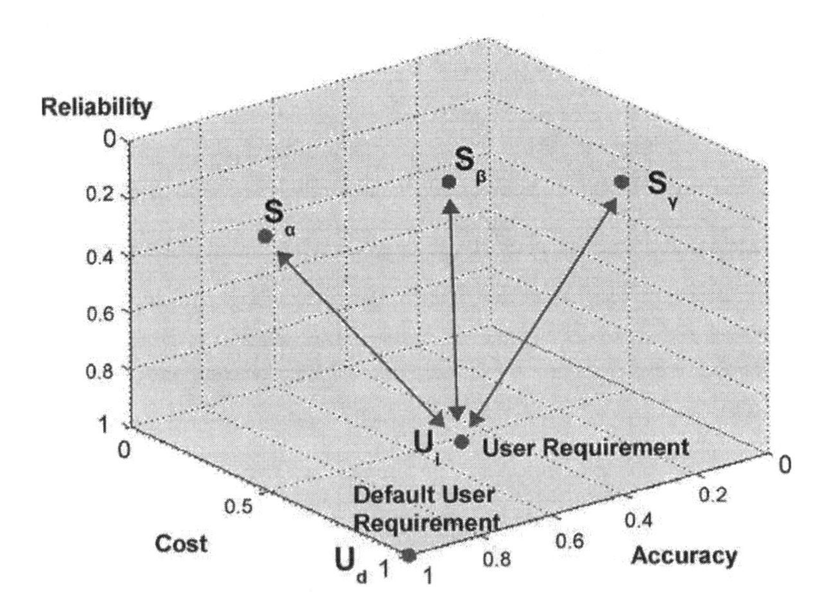

Figure 5.4: Sensors plotted in three-dimensional space for demonstration purposes. S_α, S_β, and S_γ represent real sensors. U_i represents the user preferred sensor. U_d represents the default user preferred sensor. CPWI calculates weighted distance between $S_{j=\alpha||\beta||\gamma}$ and $U_{i||d}$. Shortest distance means sensor will rank higher because it is close to the user requirement.

Algorithm 4 Comparative-Priority Based Weighted Index

Require: $(\mathbb{P}^w)$, $(\mathbb{CP})$, $(\mathbb{S}^{Indexed})$, $(\mathbb{P}^{S_j})$, $(\mathbb{UI})$
1: **Output:** $\mathbb{S}_{Ranked}$
2: $CP^{ideal} \leftarrow$ proximity based requirements$(\mathbb{UI})$
3: plot on multi-dimensional space(CP^{ideal})
4: **for** each sensor $S_j \in \mathbb{S}$ **do**
5: plot on multi-dimensional space(CP^{S_j})
6: **end for**
7: Indexing Formula (for S^{α}) $= \sqrt{\sum_{i=1}^{n}\left[W_i(U_i^d - S_i^{\alpha})^2\right]}$
8: **for** each sensor $s^j \in \mathbb{S}$ **do**
9: $\mathbb{S}^{Indexed} \leftarrow$ calculate index$(\mathbb{P}^{S_j}, \mathbb{P}^w)$
10: **end for**
11: $\mathbb{S}_{Ranked} \leftarrow$ reversed normalized ranking* $(\mathbb{S}^{Indexed})$ *i.e., lowest value is ranked higher which represents the weighted distance between use preferred sensor and the real sensors
12: **return** $\mathbb{S}_{Ranked}$

5.4.6 Context Framework

After evaluating a number of research efforts conducted in the quality of service domain relating to Web services [144], mobile computing [145], mobile data collection [67], and sensor ontologies [122], we extracted the following context properties to be stored and maintained in connection with each sensor. This information helps to decide which sensor is to be used in a given situation. We adopt the following definition of *context* for this chapter. "Context is any information that can be used to characterize the situation of an entity. An entity is a person, place, or object that is considered relevant to the interaction between a user and an application, including the user and applications themselves"[146]. CASSARAM has no limitations on the number of context properties that can be used. More context information can be added to the following list as necessary. Our context framework comprises availability, accuracy, reliability, response time, frequency, sensitivity, measurement range, selectivity, precision, latency, drift, resolution, detection limit, operating power range, system (sensor) lifetime, battery life, security, accessibility, robustness, exception handling, interoperability, configurability, user satisfaction rating, capacity, throughput, cost of data transmission, cost of data generation, data ownership cost, bandwidth, and trust.

5.5 Improving Efficiency

In this section, we present three approaches that improve the efficiency and the capability of CASSARAM. First, we propose a heuristic approach that can handle a massive number of sensors by trading off with accuracy. Second, we propose a relational-expression based filtering technique that saves computational resources. Third, we tackle the challenge of distributed sensor search and selection.

5.5.1 Comparative-Priority Based Heuristic Filtering (CPHF)

The solution we discussed so far works well with a small number of sensors. However, the model becomes inefficient when the number of sensors available to search increases. Let us consider an example to identify the inefficiency. Assume we have access to one million sensors. A user wants to select 1,000 sensors out of them. In such a situation, CASSARAM will index and rank one million sensors using proximity-based requirements provided by the user and select the top 1,000 sensors. However, indexing and ranking all possible sensors (in this case one million) is inefficient and wastes a significant amount of computational resources. Furthermore, CASSARAM will not be able to process large numbers of user queries due to such inefficiency. We propose a technique called Comparative-Priority Based Heuristic Filtering (CPHF) to make CASSARAM more efficient. The execution process is explained in Algorithm 5. The basic idea is to remove sensors that are positioned far away from the user defined ideal sensor and reduce the number of sensors that need to be indexed and ranked. Figure 5.5 illustrates the CPHF approach with a sample scenario. The CPHF approach can be explained as follows. First, all the eligible sensors are ranked in descending order of the highest weighted context property (in this case accuracy). Then, 40% (from $N_{Removable}$) of the sensors from the bottom of the list need to be removed. Next, the remaining sensors need to be ordered in descending order of the next highest weighted

Algorithm 5 Comparative-Priority Based Heuristic Filtering

Require: $(\mathbb{O})$, $(\mathbb{P})$, $(\mathbb{Q})$, (N), $(M\%)$

1: **Output:** $\mathbb{S}_{Filtered}$
2: $\mathbb{S} \leftarrow$ query ontology$(\mathbb{O}, \mathbb{Q})$
3: $\mathbb{P}^{w} \leftarrow$ get weighted priorities$(\mathbb{P})$
4: $\mathbb{P}^{Percentages} \leftarrow$ convert weights to percentages$(\mathbb{P}^{w})$
5: $N_{All} \leftarrow$ total numberof available sensors$(\mathbb{O}, \mathbb{Q})$
6: $N \leftarrow$ required number of sensors$(\mathbb{UI})$
7: $N_{Removable} \leftarrow (N_{All} - N)$
8: $\mathbb{P}^{Percentages}_{ordered} \leftarrow$ descending order$(\mathbb{P}^{Percentages})$
9: **for** each priority percentage $p \in \mathbb{P}^{Percentages}_{ordered}$ **do**
10: $\quad \mathbb{S}_{Filtered} \leftarrow$ Query $\mathbb{S}^{Filtered}$ and ordered by p
11: $\quad$ Remove $N_{Removable} \times (100 - M)$ sensors from bottom.
12: **end for**
13: **return** $\mathbb{S}_{Filtered}$

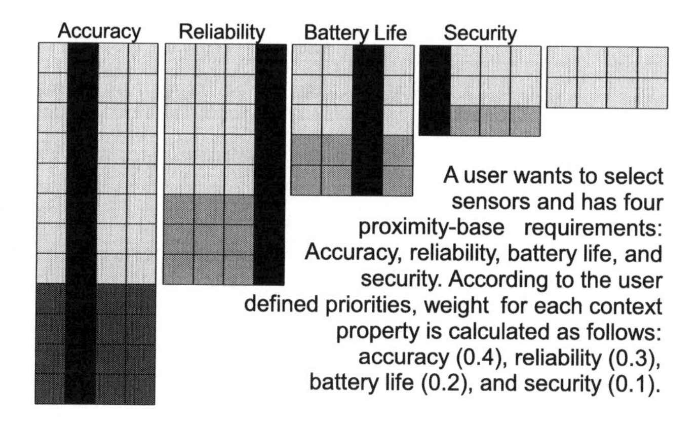

Figure 5.5: Visual illustration of Comparative-Priority Based Heuristic Filtering.

context property (in this case reliability). Then, 30% (from $N_{Removable}$) of the sensors from the bottom of the list need to be removed. This process needs to be applied for the remaining context properties as well. Finally, the remaining sensors need to be indexed and ranked. This approach dramatically reduces the indexing and ranking related inefficiencies. Broadly, this category of techniques is called *Top-K selection* where top sensors are selected in each iteration. The efficiency of this approach is evaluated and discussed in Section 5.7.

5.5.2 Relational-Expression Based Filtering (REF)

This section explains how computational resources can be saved and how to speed up the sensor search and selection process by allowing the users to define preferred context property values using relational operators such as $<, >, \leq$, and $\geq$. For example, users can define an upper bound, lower bound, or both, using relational operators. All context properties defined by relational operators, other than the equals sign (=), are considered to be semi-non-negotiable requirements. According to CASSARAM, non-negotiable as well as semi-non-negotiable requirements are defined using semantic queries. Let us consider a scenario where a user wants to select sensors that have 85% accuracy. However, the user can be satisfied by providing sensors with accuracy between 70% and 90%. Such requirements are called semi-non-negotiable requirements. Defining such a range helps to ignore irrelevant sensors during the semantic querying phase without even retrieving them to the CPWI generating phase, and this saves computational resources. Even though users may define ranges, the sensors will be ranked considering the user's priorities by applying the same concepts and rules as explained in Section 5.4. The efficiency of this approach is evaluated in Section 5.7.

5.5.3 Distributed Sensor Searching

We have explained how CASSARAM works in an isolated environment without taking into consideration the distributed nature of the problem. Ideally, we expect that not all sensors will be connected to one single server (e.g., a single middleware instance). Similarly, it is extremely inefficient to store complete sensor descriptions and related context information in many different servers in a redundant way. Ideally, each CPS middleware instance should keep track of the sensors that are specifically connected to them. This means that each server knows only about a certain number of sensors. However, in order to deal with complex user requirements, CASSARAM may need to query multiple CPS middleware instances to search and select the suitable sensors. Let us consider a scenario related to the smart agriculture domain [141]. A scientist wants to find out whether his experimental crops have been infected with a disease. His experimental crops are planted in fields distributed across different geographical locations in Australia. Furthermore, the sensors deployed in the fields are connected to different CPS middleware instances, depending on the geographical location. In order to help the user to find the appropriate sensors, CASSARAM needs to query different servers in a distributed manner. We explored the possibilities of performing such distributed queries efficiently. We identified three different ways to search sensors distributively, depending on how the query/data would be transferred over the network (i.e., path), as depicted in Figure 5.6. We also identified their strengths, weaknesses, and applicability to different situations.

1) Chain Processing: Data is sent from one node to another sequentially as depicted in Figure 5.6(a). First, a user defines his requirements using a CPS middleware instance (e.g., GSN installed in a particular server). Then, this server becomes the search request initiator (SRI) for that specific user request. The SRI processes the request and selects the 100 most appropriate sensors. Then, the information related selected sensors (i.e., the unique IDs of the sensors and respective CPWIs) is sent to the next server node. The second node (i.e., that next node) merges the incoming sensor information with the

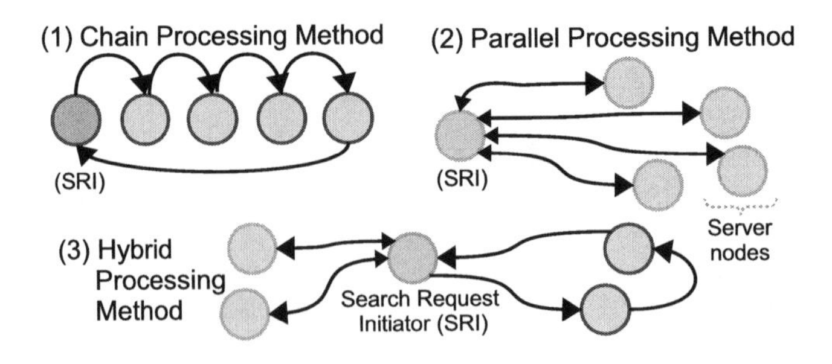

Figure 5.6: Distributed processing approaches for CASSARAM.

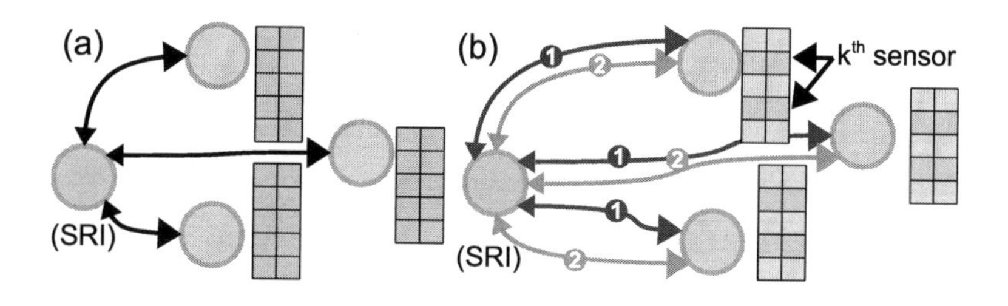

Figure 5.7: Optimization: (a) wihout k-extension and (b) with k-extension.

existing sensor descriptions and performs the sensor selection algorithm and selects the 100 best sensors. This pattern continues until the sensor request has visited all the server nodes. This method saves communication bandwidth by transferring only the most essential and minimum amount of data. In contrast, due to a lack of parallel processing, the response time could be high.

2) Parallel Processing: The SRI parallelly sends each user search request to all available nodes. Then, each sensor node performs the sensor searching algorithm at the same time. Each node selects the 100 most appropriate sensors and returns the information related selected sensors to the SRI. In circumstances where we have 2500 server nodes, the amount of data (2500×100) received by the SRI could be overwhelming, which would waste the communication bandwidth. The SRI processes the sensor information (2500×100) and selects the final 100 most appropriate sensors. This approach becomes inefficient when N becomes larger.

3) Hybrid Processing: By observing the characteristics of the previous two methods, it is obvious that the optimal distributed processing strategy should employ both chain and parallel processing techniques. There is no single method that works efficiently for all types of situations. An ideal distributed processing strategy for each situation needs to be designed and configured dynamically depending on the context, such as the types of the devices, their capabilities, bandwidth available, and so on.

We can improve the efficiency of the above methods as follows. In the parallel processing method, each node sends information related to N sensors to the SRI as depicted in Figure 5.7(a). However, at the end, the SRI may only select N sensors (in total) despite its having received a significant amount of sensor related information $(N \times number\ of\ nodes)$. Therefore, the rest of the data $[(N \times number\ of\ nodes) - N]$ received by the SRI would be wasted. For example, let us assume that a user wants to select 10,000 sensors. Assuming that there are 2500 server nodes, the SRI may receive a significant amount of sensor information $(10,000 \times 2500)$. However, it may finally select only 10,000 sensors. We propose the following method to reduce this wastage, depicted in Figure 5.7(b).

In this method, the SRI forwards the search request to each server node parallelly, as depicted in step (1) in Figure 5.7b. Each node selects the 10,000 most appropriate sensors. Without sending information about these 10,000 sensors to the SRI, each server node sends only information about the kth sensor (the UID and CPWI of every kth sensor). (That is, if $k = 1,000$, then the server node sends only the 1000th, 2000th, 3000th, ... 10,000th sensors.) Therefore, instead of sending 10,000 records, now each server node returns only 10 records. Once the SRI receives the sensor information from all the server nodes, it processes and decides which portions need to be retrieved. Then, the SRI sends requests back to the server nodes and now each node returns the exact portion specified by the SRI (e.g., the 5th server node may return only the first 2000 sensors instead of sending 10,000 sensors) as depicted in (2). In this method, k plays a key role and has a direct impact on the efficiency. k needs to be chosen by considering N as well as other relevant context information as mentioned earlier. For example, if we use a smaller k, then information about more sensors would be sent to the SRI during step (1), but with less wastage in step (2). In contrast, if we use a larger k, then less information would be sent to the SRI during step (1), but there would be comparatively more wastage in step (2). Furthermore, machine learning techniques can be used to customize the value of k for each server node, depending on the user's request and context information, such as the types of the sensors, energy, bandwidth availability, etc. The suitability of each approach is discussed in Section 5.7.2.

5.6 Implementation and Experimentation

In this section, we describe the experimental setup, datasets used, and assumptions. The experimental scenarios we used are explained at the end. The discussions related to the experiments are presented in Section 5.7.

We analyzed and evaluated the proposed model using a prototype called *CASSARA Tool*, which we developed using Java. The data was stored in a MySQL database. Our tool allows capturing user preferences and the priorities of the various context properties of the sensors. We used a computer with an Intel(R) Core i5-2557M 1.70GHz CPU and 4GB RAM to evaluate our proposed model. We also reproduced the experimentations using a higher-end computer with more CPU and RAM and the results showed that the graphs are similar in shape though the exact values are different. In order to perform mathematical operations such as a Euclidean distance calculation in multi-dimensional space, we used the Apache Commons mathematics [147] library. It is an open-source optimized library of lightweight, self-contained mathematics and statistics components, addressing the most common problems not available in the Java programming language. As we used a Semantic Sensor Ontology (SSN) [122] to manage the sensor descriptions and related data, we employed open-source Apache Jena API [148] to process and manipulate the semantic data. Our evaluation used a combination of real data and synthetically generated data. We collected environmental linked data from the

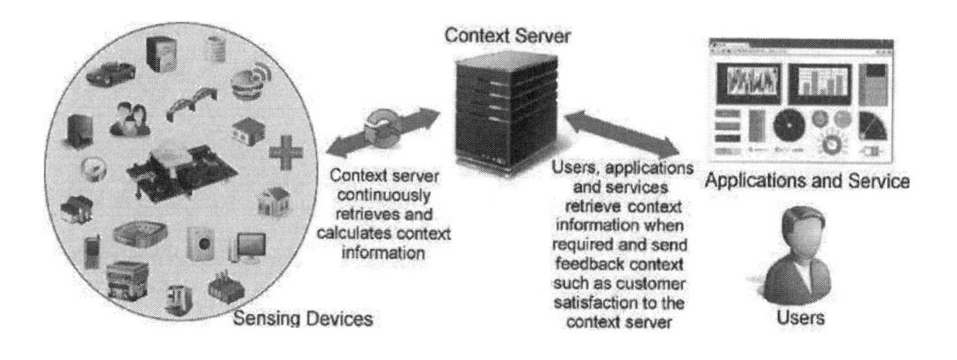

Figure 5.8: Context information collection and modelling.

Bureau of Meteorology [149] and datasets from both the Phenonet project [142] and the Linked Sensor Middleware (LSM) project [127, 128].

The main reasons for combining the data were the need for a large amount of data and the need to control different aspects (e.g., the context information related to the sensors needed to be embedded into the dataset, because real data that matches our context framework is not available in any public data sets at the moment) to better understand the behavior of CASSARAM in different CPS related real-world situations and scenarios where real data is not available. We make the following assumptions in our work. We assume that the sensor descriptions and context information related to the sensors have already been retrieved from the sensor manufacturers in terms of ontologies, and been used in the SSN ontology as depicted in Figure 5.8. Similarly, we assume that the context data related to the sensors, such as accuracy, reliability, etc., have been continually monitored, measured, managed, and stored in the SSN ontology by the software systems. In order to evaluate the distributed processing techniques, we proposed an experimental test involving four computational nodes. All the nodes are connected to a private organizational network (i.e., The Australian National University IT Network). The hardware configurations of the three additional devices are as follows: (1) Intel Core i7 CPU with 6GB RAM, (2) Intel Core i5 CPU with 4GB, and (3) Intel Core i7 with 4G. The details are presented in Section 5.7.2.

5.7 Performance Evaluation

We evaluated CASSARAM using different methods and parameters as depicted in Figures 5.9(a)–5.9(i). In this section, we explain the evaluation criteria which we used for each experiment and discuss the lessons we learned. Figure 5.9(a) shows how the storage requirement varies depending on the number of sensor descriptions. We stored the data according to the SSN ontology,

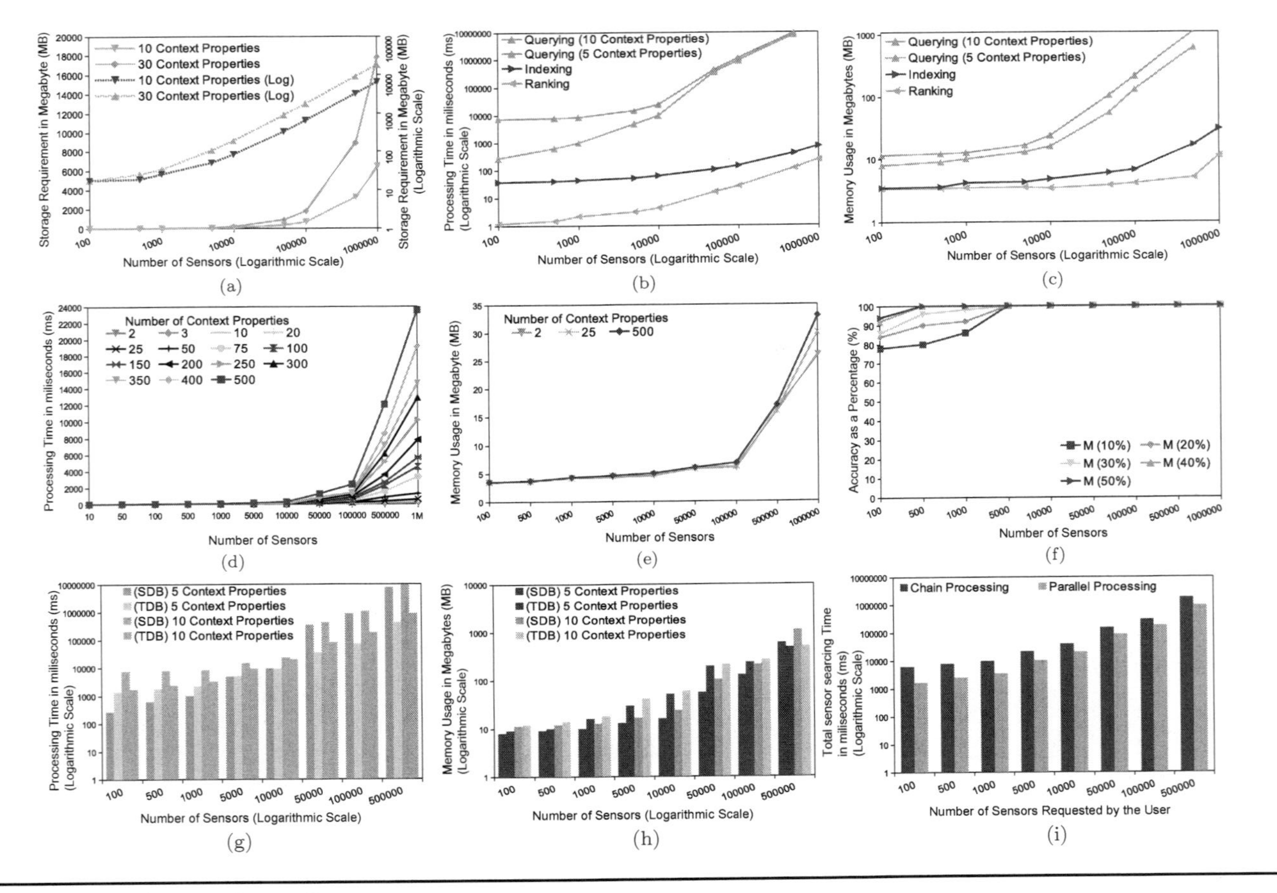

Figure 5.9: Experimental results.

as depicted in Figure 5.3. We conducted two experiments where we stored 10 context properties and 30 context properties from the context framework we proposed in Section 5.4.6. To store one million sensor descriptions, it took 6.4 GB (10 context properties) and 17.8 GB (30 context properties). It is evident that the storage requirements are correlated with the number of triples: a single triple requires about 0.193 KB storage space (for 100,000+ sensors). Though storage hardware is becoming cheaper and available in high capacities, the number of context properties that need to be stored should be decided carefully in order to minimize the storage requirements, especially when the number of sensors is in the billions.

Figure 5.9(b) shows how much time it takes to select sensors as the number of sensors increases. Each step (i.e., searching, indexing, and ranking) has been measured separately. Semantic querying requires significantly more processing time than indexing and ranking. Furthermore, as the number of context properties retrieved by a query increases, the execution time also increases significantly. Furthermore, it is important to note that MySQL can join only 61 tables, which only allows retrieving a maximum of 10 context properties from the SSN ontology data model. Using alternative data storage or running multiple queries can be used to overcome this problem. Similarly, it is much more efficient to run multiple queries than to run a single query if the number of sensors is less than 10,000 (e.g., 8 ms to retrieve 5 context properties and 24 ms to retrieve 10 context properties when querying 10,000 sensors). In addition, Figure 5.9(c) shows how much memory is required to select sensors as the number of sensors increases. It is evident that having more context properties requires having more memory. The memory requirements for querying do not change much up to 10,000 (ranging from 10 MB to 25 MB). When the number of sensors exceeds 10,000, the memory requirements grow steadily, correlated with the number of sensors. In comparison, indexing and ranking require less memory.

Figure 5.9(d) shows the processing time taken by the sensor indexing process as the number of context properties and the number of sensors increase. Reducing the number of sensors needing to be indexed below 10,000 allows speeding up CASSARAM. The processing time starts to increase significantly after 100,000 sensors. Similarly, Figure 5.9(e) shows the memory usage by the sensor indexing process as the number of context properties and sensors increases. Even though the memory requirements increase slightly, the actual increase is negligible when the number of sensors is still less than 100,000. After that, the memory requirements increase substantially, but are still very small compared to the computational capabilities of the latest hardware. Furthermore, the number of context properties involved does not have any considerable impact during the indexing process. The differences only become visible when the number of sensors reaches one million. Still, the memory required by the process is 30 MB. Java garbage collection performs its task more actively when processing large numbers of sensors, which makes the difference invisible.

Figure 5.9(f) shows how the accuracy changes when the margin of error (M%) value changes in the CPHF algorithm and the number of sensors increases. The scenario presented in Figure 5.5 has been evaluated. The accuracy of the CPHF approach increases when the margin of error (M) increases. However, a lower M leads CASSARAM toward low resource consumption. Therefore, there is a trade-off between accuracy and resource consumption. The optimum value of M can be dynamically learned by machine learning techniques based on which context properties are prioritized by the users in each situation and how the normalized weights are distributed between the context properties.

5.7.1 Evaluating Alternative Storage Options

In the evaluations conducted earlier (Figures 5.9(g) to 5.9(i)), we used Jena SDB/MySQL-backed RDF storage to store the data. In order to evaluate the performance of CASSARAM when using alternative storage options, we here employ a Jena TDB-backed approach (jena.apache.org/documentation/tdb). In Figure 5.9(g), we compare the processing times taken by both the Jena SDB/MySQL and the Jena TDB approach. Furthermore, in Figure 5.9(h), we compare the memory usage by the SDB and TDB approaches. According to the Berlin SPARQL Benchmark [143], Jena TDB is much faster than Jena SDB. We also observed similar results both in 5 context data processing as well as in 10 context data processing. Specifically, Jena TDB is 10 times faster than SDB when processing 10 context properties, where the dataset consists of half a million sensor descriptions. The Jena SDB approach consumed less memory than the Jena TDB approach when the dataset was less than 100,000 sensor descriptions. However, after that, the Jena TDB approach consumes less memory than the Jena SDB. Specifically, Jena TDB uses 50% less memory than Jena SDB when processing 10 context properties, where the dataset consists of half a million sensor descriptions. Therefore it is evident that Jena TDB is more suitable when the number of sensor descriptions goes beyond 100,000.

Despite the differences we observed in our evaluation, there are several factors that need to be considered when selecting underlying storage solutions. As evaluated on the Berlin SPARQL Benchmark, there are several other storage options available, such as Sesame (openrdf.org), Virtuoso TS, Virtuoso RV, and D2R Server [143]. Jena TDB offers faster load times and better scale, but has the worst query performance. Sesame seems better all-round for low data sizes assuming infrequent loads. In contrast, Jena SDB provides moderate performance, offering load times, query performance, and scalability between the Jena TDB and Sesame. Based on these evaluations, at the time at which this chapter was written, there is no superior solution that has all good qualities. Due to the lack of extensive usage and the short existence of Sesame, SDB/MySQL can be seen as a better choice especially when considering database functionalities such as backup, concurrent and parallel processing. As we do not expect frequent loading/ unloading of datasets such as sensor descriptions, it is evident that SDB outperforms TDB in query processing (excluding data loading) [143]. As we expect more updates (transactions) to occur, SDB would be a better choice.

5.7.2 Evaluating Distributed Sensor Searching

We evaluated distributed sensor searching using a private network that consists of four computational nodes. We compare two different distributed sensor search techniques, namely, chain processing and parallel processing with/without k-extensions, which we discussed in Section 5.5.3. The results are presented in Figure 5.9(i). Each node consists of a dataset of one million sensor data descriptions. The four datasets are different from each other. Five context properties are considered for the evaluation and the context information is stored using Jena TDB. First, we discuss the techniques from the theoretical perspective.

Let us define some of the notations which will be used in the following discussion: n= number of computational nodes (in our experiments n=4), N=number of sensors requested by the users, S_i= number of sensor descriptions stored in the ith computational node, r= size of a single sensor description record (i.e., storage requirements), $t_{i,j}^{net}$= time taken for network communication between the computational nodes i and j, t_i^{pro}= time taken to query the computational node i; merge the indexed results with the incoming results, and select the final number N of sensors. The total time taken by chain-based distributed sensor searching can be defined as:

$$Total_{chain} = \sum_{i=1}^{n} t_i^{pro} + \sum_{i=1}^{n-1} t_{i,i+1}^{net} + t_{n,1}^{net} \tag{5.1}$$

The total time taken by parallel distributed sensor searching can be defined as:

$$Total_{parallel} = max\left\{i = [2..n] : t_i^{pro} + t_{1,i}^{net}\right\} \tag{5.2}$$

According to the results, it is evident that parallel processing is more efficient than chain processing in terms of the total processing time. However, parallel processing is inefficient in other aspects, such as network communication and bandwidth consumption. Therefore, we proposed k-extension to address this issue. The evaluation of the k-extension approach is presented in Table 5.2. In this experiment, we measured how much data communication can be saved (i.e., due to elimination of redundant data communication that occurs in parallel processing without k-extension) by using different k values under different N values. We measured the *guaranteed minimum*[2] amount of data communications (measured in Megabytes) that can be saved.

In Table 5.2, positive values indicate the minimum amount of data communication saved using the k-extension. Although negative values indicate no

[2]Depending on the dataset and the context information stored in each node, the parallel processing technique with k-extension will be able to save more data communication than the guaranteed minimum level.

Table 5.2: The Amount of Redundant Data Communication Saved by the Parallel Sensor Search with k-extension Strategy

k value	Number of sensors requested by the users (N)								
	100	500	1,000	5,000	10,000	50,000	100,000	500,000	1,000,000
10	-60.7	-60.5	-60.3	-58.7	-56.7	-40.5	-20.2	141.6	344.0
100		-5.9	-5.7	-4.1	-2.1	14.1	34.3	196.2	398.5
500			-1.1	0.5	2.5	18.7	38.9	200.8	403.1
1000				0.8	2.8	19.0	39.3	201.1	403.5
5000					0.9	17.1	37.3	199.2	401.5
10000						14.1	34.3	196.2	398.5
50000							10.1	172.0	374.3
100000								141.6	344.0
500000									101.2

in Megabytes (MB)

guaranteed savings, some situations have a high chance of saving redundant data communication compared to others. Equation (5.3) can be used to calculate the guaranteed minimum amount of data saving by using k- extensions.

$$Total_{\text{Saving}} = \sum_{i=2}^{n} S_i r - \left\{ [\sum_{i=2}^{n} \frac{S_i}{k} + N + (k-1)n] \times r \right\}, \text{ IF } (k < N) \quad (5.3)$$

Let us consider different scenarios where chain and parallel processing can be used. Chain processing is suitable for situations where saving computational resources and bandwidth is more critical than response time. A parallel processing method *without k-extension* is suitable when response time is critical and N is fairly small. *k-extension* requires two communication rounds: communication radios need to be opened and closed twice. Such a communication pattern consumes more energy [62], especially if the computational devices are energy constrained. Therefore, transmitting data all at once is more efficient. However, this recommendation becomes invalid when N becomes very large (10,000+). Our experiments clearly show that k-extensions can be used to improve the efficiency of the parallel sensor searching approach, especially when N is large. The ideal value of k needs to be determined based on N, n, and S_i.

5.8 Open Challenges and Future Research Directions

In this section, we identify a number of major challenges that need to be addressed. First, we discuss the challenge of collecting and modelling context information related to sensors. Then, we explain the notion of providing access to sensors as a service and the related challenges. Thirdly, we highlight the importance of automated configuration of sensors and the challenges involved in context information management in highly dynamic environments. Finally, we elaborate the role of context information on developing optimized sensing strategies.

5.8.1 Context Discovery, Processing, and Storage

Throughout this chapter, we assumed that context information related to sensors has been collected and modelled in semantically annotated data and is ready to be used. However, in the real world such context information needs to be collected from each Internet-connected object. Further, collecting such context data is challenging. Some context information such as battery life can be directly retrieved from the objects. However, context information such as expected lifetime may need to be retrieved from data sources provided by the object manufacturer or technical documents. Another set of context information such as accuracy or reliability needs to be generated by fusing (e.g., pattern recognition) data collected for some period. Furthermore, separate context information profiles need to be maintained for each sensor. Such an application programming interface (API) that supports collecting and modelling context information is yet to be developed. In this chapter, we modelled context information using semantic modelling technologies. The data storages we used are TDB and SDB. It is important to consider and evaluate the state-of-the-art data storage technologies such as NOSQL data store [150]. Further, efficiency of querying can be improved by applying techniques such as MapReduce [151].

5.8.2 Utility Computing Models and Sensing as a Service

In this chapter, we did not calculate the costs in financial terms. In utility based cloud computing, CPU, RAM, and data storage per hour are sold in pay-as-you-go fashion. In sensing as a service model, sensors (or sensor data) also need to be valued (provided a price tag) in addition to the above-mentioned computational resources. Such valuation has not yet been addressed by the research community. There are interesting factors that need to be considered when valuing sensor data. Factors such as battery life and location may impact the value of the sensor data. For example, smart devices that have continuous power supply (e.g., indoor temperature sensors) may be cheaper in comparison to the devices with limited and fixed power (e.g., temperature sensors deployed in a forest) due to cost of recharging or disposable nature. If the cloud service provider performs additional analytics, additional charges may apply.

5.8.3 Automated Smart Device Configuration

Throughout this chapter, we assumed smart devices are connected to the cloud, and configuration has been performed. However, detecting, discovering, and configuring smart objects is a tough challenge. This is especially challenging due to dynamicity where smart devices may appear and disappear at a given location over time. It is vital to maintain a context information profile of each device even when it is not at the given location. The MAC address can be used uniquely to identify the smart device. However, some of the context information may change drastically depending on the location of the smart device. For example, a smart object that has a pressure sensor may perform well when deployed inside a shopping mall. However, if we deploy the same object outside the shopping mall, accuracy may drastically change due to natural factors such as temperature, wind, rain, and so on. So it becomes challenging to maintain location-aware context profiles. In addition to the location, time of the day, day of year, seasonal variation may have a significant impact on context information such as reliability. Therefore, an ideal solution may be to model context information by considering the above-mentioned factors. Preliminary efforts on automated smart device configuration are presented in [64].

5.8.4 Optimize Sensing Strategy Development

Scheduling sensor networks has been a widely researched topic. However, Internet connected devices and their scheduling creates a different kind of challenge. Let us consider a scenario related to smart agriculture to understand why context-aware scheduling matters in sensor configuration. Severe frosts and heat events can have a devastating effect on crops. Flowering time is critical for cereal crops and a frost event could damage the flowering mechanism of the plant. However, the ideal sampling rate could vary depending on both the season of the year and the time of day. For example, a higher sampling rate is necessary during the winter and the night. In contrast, lower sampling would be sufficient during the summer and daytime. On the other hand, some reasoning approaches may require multiple sensor data readings. For example, a frost event can be detected by fusing air temperature, soil temperature, and humidity data. However, if the air temperature sensor stops sensing due to a malfunction, there is no value in sensing humidity because frost events cannot be detected without temperature. In such circumstances, configuring the humidity sensor to sleep is ideal until the temperature sensor is replaced and starts sensing again. Such intelligent (re-)configuration can save energy by eliminating ineffectual sensing and network communication.

5.9 Summary

Device searching is an important functionality in the cyber-physical systems domain. Search may directly help the users or may be useful as a subroutine of a complex process (e.g., on demand sensors and services composition [152]). This chapter explains how the device search fits in the Internet of Things architecture. We discussed different types of searching techniques. Multiple searching techniques can be combined in order to build a complete CPS middleware platform. Some techniques are less resource intensive where others are more resource intensive. However, it is important to note that such combinations will increase the computational complexity. We presented a context-aware sensor search, selection, and ranking model, called CASSARAM, to address the challenge of efficiently selecting a subset of relevant sensors out of a large set of sensors with similar functionality and capabilities.

We showed how the context information can be used to search the sensors that are best suited to user requirements in sensing as a service scenario. We selected sensors based on the user's expectations and priorities. As a proof of concept, we built a working prototype to demonstrate the functionality of CASSARAM. We also highlighted how CASSARAM helps to achieve the broader vision of sensing-as-a-service in the CPS paradigm. CASSARAM allows optimizing the sensor data collection approaches by selecting the sensors in an optimized fashion. For example, CASSARAM can be used to find sensors which have more energy and collect data only from those sensors. Such a strategy helps to run the entire sensor network for a much longer time without reconfiguring. We explored three different techniques that improve the efficiency and scalability of CASSARAM: comparative-priority based heuristic filtering, relational-expression based filtering, and distributed sensor searching. Finally, it is important to mentioned that the goal of building a comprehensive smart device searching platform is yet to be achieved by the research community. Addressing the open challenges mentioned previously will help to move in that direction.

Chapter 6

Energy Management for CPS

Chi Harold Liu

Beijing Institute of Technology, China

CONTENTS

6.1 Introduction

The previous three chapters address different aspects of the architectures of CPS. However, there are various technical challenges in sensor energy and data quality management of CPS. A major one that drives our work involves the large-scale management of heterogeneous devices that are expected to populate CPS systems. A great many sensor types, manufacturers, protocols, etc., are expected to co-exist and hence, any solution must be designed to operate as expected regardless of the device configuration. Regarding energy management, this motivates the need for a universal management approach that attempts to control MAC (medium access control) level energy consumption of nodes, as motivated by previous research [153]. Furthermore, an efficient management scheme should minimize the transmission of control messages crossing different domains, and thus we are seeking a *long-term* optimal solution. In regard to data quality management, a universal measure of expression can be found in recent work in quality-of-information (QoI) management. Broadly speaking, QoI relates to the ability to judge whether information is *fit-for-use* for a particular purpose [154, 155, 156]. For the purposes of this chapter, we will assume that QoI is characterized by a number of attributes including accuracy, latency, and physical context (specifically, sensor coverage in this chapter [154]).

To address the aforementioned challenges, we aim to design an energy management service (and supporting algorithms) that is transparent to and compatible with any lower layer protocols and over-arching applications, while providing *long-term* energy-efficiency under the satisfactory QoI constraints. In support of our design, we first introduce the new concept of "sensor-to-task relevancy" to explicitly consider the sensing capabilities offered by a sensor (or a set thereof) to the applications and QoI requirements required by a task. Second, we use the generic information fusion function to compute the "critical covering set" of any given task in selecting the sensors to service a task over time. Third, we propose a runtime energy management framework based on the previous design elements to control the duty cycles of a sensor in the long run, i.e., the control decision is made optimally considering the long-term task usage statistics where the service delay of each task serves as the constraint. Then, an extensive case study related to water quality monitoring is given to demonstrate the ideas and algorithms proposed in this chapter, and a simulation is made to support all performance analysis. Lastly, we further consider the signal transmission and processing latency into our previous proposal to both theoretically and experimentally investigate its impact on the measured delay probability. Finally, we provide some potential implementation guidelines to make the energy management framework more applicable under realistic scenarios. It should be noted that this is first-of-a-kind research that manages the energy usage of a variety of sensors from different domains, irrespective of how the provided sensing capabilities will be used by different applications.

The rest of this chapter is organized as follows. Section 6.2 presents related research efforts. The system model, including the system flow of the proposed efficient energy management framework, is described in Section 6.3. The sensor-to-task relevancy and the critical covering set are introduced in Section 6.4, and the optimization problem of efficient energy management is formulated in Section 6.5, where several solutions are also given and analyzed. In Section 6.6, a case study of water quality monitoring is explained in detail, and its simulation results are presented. In Section 6.7 the modeling of signaling overhead is thoroughly discussed and analyzed. Implementation guidelines are given in Section 6.8. Finally, concluding remarks are drawn in Section 6.9. This chapter largely extends [157], by introducing the delay model for all tasks in a probabilistic manner (see Section 6.5.2), giving the explicit definition of the weight factor in the duty cycle optimization (see Section 6.5.3), demonstrating extensive performance evaluation results (see Section 6.6), and adding the modeling and analysis of signal transmission and processing latency (see Section 6.7).

A summary of important symbols used in this chapter is listed in Tables 6.1 and 6.2.

6.2 Related Work

The area of QoI has been proposed recently to judge how retrieved information is fit-for-use for a particular task [158, 159]. Work described in [160] made further contributions to this area by proposing a QoI satisfaction index to describe the degree of QoI satisfaction that tasks receive from a wireless sensor network (WSN). Reference [160] additionally describes a QoI network capacity metric to describe the marginal ability of the WSN to support the QoI requirement of a new task upon its arrival to the network. Based on these, the authors proposed an adaptive admission control scheme to optimally accommodate the QoI requirements of all incoming tasks by treating the whole WSN as a "black box."

Existing work describes many different schemes for WSN node scheduling [161, 162, 163, 164, 165]. In [161], Ma et al. propose centralized and distributed algorithms to assign sensors with consecutive time slots to reduce the frequency of operational state transitions. In [162], the authors describe an energy-efficient scheduling scheme for low-data-rate WSNs, where both the

Table 6.1: Summary of Important Symbols

Symbol	Definition (Section where the symbol is first used)
$\underline{c}_n$	sensing capability of sensor n (6.3.1)
E_n	initial energy reserve of sensor n (6.3.1)
$\bar{E}_n(t)$	residual energy of sensor n at time t (6.3.1)
P_n^{on}	power consumption level of sensor n when it is ON (6.3.1)
$\underline{q}_m$	desired QoI of task m (6.3.2)
L	frame length (6.3.3)
r_{nm}	relevancy of sensor n to task m (6.4)
$f(\cdot)$	sensor-to-task relevancy function (6.4)
$g(\cdot)$	information fusion function (6.4.1)
T	lifetime of the CPS sensory environment (6.5.1)
$\eta_n(t)$	runtime generalized duty cycle of sensor n up to t (6.5.1)
$N_n(t)$	number of switches sensor n makes up to time t (6.5.1)
$\Sigma_n(t)$	aggregation of the ON times of sensor n up to time t (6.5.1)
P_n^{sw}	energy consumed each time sensor n switches mode (6.5.1)
$d_{m,i}$	service delay of task m for its i-th instance (6.5.2)
π_m	steady state probability of task m (6.5.2)

sensors' energy requirements for specific operational states and the state transitions are considered. In [163], the authors propose a cross-layer framework to optimize global power consumption and balance the load in sensor networks. In [164], the authors utilize control theory to balance energy consumption and packet delivery success. In [165], the authors present novel schemes to reduce sleep latency, while achieving balanced energy consumption among sensor nodes with energy harvesting capability. The authors in [166] study the QoI performance of WSN tracking applications with energy constraints, focusing on a duty-cycled network with random wake-up schedules for different nodes. In comparison, our work is different in that we explicitly consider the multi-dimensional QoI requirements of tasks and capabilities of sensors, and use this in addition to energy to determine node duty cycle schedules. This approach completely decouples relations between sensors and applications, but dynamically controls the energy consumption state of each sensor to achieve desired QoI over the long run.

Finally, we note that there has been plenty of work on MAC layer protocol design for WSNs focusing on minimizing energy consumption in order to achieve prolonged system lifetime, such as S-MAC [167] and T-MAC [168]. In contrast to this work, our proposal is a system-level management operation and not a communications protocol; and more importantly, this proposal can work with systems that engage to MAC level energy conservation as well.

Table 6.2: Summary of Important Symbols (continued)

Symbol	Definition (Section where the symbol is first used)
$\zeta_{m,i}$	delay failure indicator of task m for its i-th instance (6.5.2)
ζ_m	average measured delay failure of task m (6.5.2)
ξ_m	delay failure threshold of task m (6.5.2)
τ_m	delay tolerance of task m (6.5.2)
β_n	weight factor in duty cycle minimization (6.5.3)
$\mathbf{P}$	task transition matrix for the exclusive task model (6.5.4)
$\Delta_n(t)$	energy consumption of sensor n between t and $t+L$ (6.5.4)
$P_m(t)$	probability that an instance of task m starts at t (6.5.4)
$\Psi_m(t)$	probability that at least one CCS of task m exists at t (6.5.4)
$\Phi_{m,i}(t)$	probability that task m's i-th instance starts at t (6.5.4)
$Q(\cdot)$	tail probability of the standard normal distribution (6.5.4)
$Z_m(t)$	fused sensory information for task m at time t (6.6.1)
$W_n(t)$	retrieved sensory information from sensor n at time t (6.6.1)
δ_m, ϵ_m	accuracy requirement of task m (6.6.1)
μ	parameter for exponential task instance duration (6.6.1)
$l_{\min}$	minimum task instance duration (6.6.1)
ω	number of delayed frames (6.7.1)
l_k	the duration of the k-th instance of all tasks (6.7.1)

6.3 System Model

In this section, we present the modeling of sensors, tasks, and overall system architecture and flow.

6.3.1 Sensors

We consider an CPS sensory environment that comprises a collection $\mathcal{N}$ of N sensors (indexed by $n \in \{1, 2, \ldots, N\}$), plus a gateway (the sink). Each sensor is associated with certain sensing, processing, and communication capabilities. The sensing capability of a sensor represents its ability to offer a certain level of QoI to a task, but independently of any specific task. The sensing capability of sensor n is described by the K-vector $\underline{c}_n \in \mathbb{R}^K$, whose entries include QoI attributes such as the measurement errors, latency in reporting, its coverage, etc. For each sensor n, the initial energy reserve is denoted as E_n, and the residual energy at time t is denoted as $\bar{E}_n(t)$.

We assume that there are only two power consumption levels for a sensor: (a) P_n^{on} when in active mode, and (b) negligibly small (relative to P_n^{on}) approximated with 0 when in sleeping mode.

6.3.2 Tasks

We consider a collection $\mathcal{M}$ of M tasks (indexed by $m \in \{1, 2, \ldots, M\}$). Each task represents a specific class of activities that may share a common spatial property but not temporal properties, such as starting time or duration. An *instance* of a task represents a single continuous period that the task is in service. For example, "monitor the water quality at certain location" may represent one of the M tasks, while doing so between times t_1 and t_2 or t_3 and t_4, represents two instances of the same sensing task executed over two different time periods. Each task's desired QoI is described by a K-vector $\underline{q}_m$, describing the desired accuracy, latency, coverage, etc. Note that the elements in $\underline{q}_m$ can be vectors as well, as a QoI requirement can be defined by more than one parameter, as illustrated in a case study in Section 6.6.1.

Finally we consider the granularity for all tasks that cannot be separated into sub-tasks. If a submitted task includes a combination of different tasks, we consider the joint set of their QoI requirements into our framework.

6.3.3 System Flow

We assume that our energy management scheme runs within the CPS EMS, interacting with both the applications and gateways of different underlying network domains, so that control signals can be computed, generated, and sent to the sensors.

We consider a discrete (or slotted) time system operation. Duty-cycling decisions are made by the EMS every L time slots, which define the duration of a L-slot *frame* in the system. Then decisions are sent to the gateways that coordinate control operations of the corresponding sensors. For simplicity, we assume a sensor stays active in a frame after it is woken up. Contrary to the collection $\mathcal{N}$, the gateway is assumed to have sufficient processing power and energy capacity. We assume that the frame length L is far less than the average service time of tasks and the average idle time between two consecutive instances of a task, which ensures that the probability of any task changing its status during a frame is negligible. We also assume that the current service time is known to the CPS EMS and gateway after it starts.

Figure 6.1 illustrates the procedure for the proposed energy management framework during one frame, which can be summarized as follows:

1. At the sensor deployment stage, compute the *critical covering sets* (CCSs) of each task with information fusion algorithms, based upon the sensor capabilities and the desired QoI of the tasks (see Section 6.4).

2. At the beginning of a frame, the EMS makes a decision on *when* to activate/deactivate which sensor and for *how long* in the current frame based on the task model and sensor status, and sends the control message back to each sensor through its gateway (see Section 6.5).

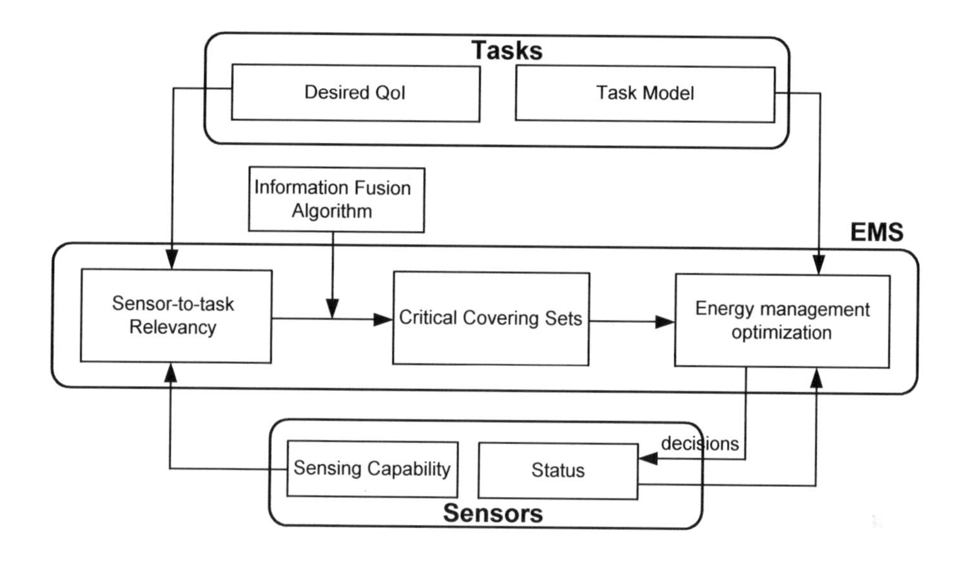

Figure 6.1: System flow of the proposed energy management framework.

3. During a frame, each sensor follows its predetermined waking-up schedule without further communications with the gateway until the next frame.

6.4 QoI-Aware Sensor-to-Task Relevancy and Critical Covering Sets

In [158], the 5WH principle was proposed to summarize the information needs of a task and the sensing capabilities of a WSN, and in [169], the spatial relevancy of the provided information was introduced along with a way to measure it. Motivated by this prior work, we propose the relevancy of a sensor to a task as the degree to which the sensor can satisfy the task's QoI requirements. Specifically, we define:

$$r_{nm} = f\left(\underline{c}_n, \underline{q}_m\right) \in [0,1], \quad \forall n \in \mathcal{N}, m \in \mathcal{M}, \tag{6.1}$$

where r_{nm} denotes the relevancy of sensor n to task m, and $f(\cdot)$ is a generic relevancy function that takes value in $[0,1]$ by definition. A specific example of f is given in Section 6.6.1.

We define a sensor *irrelevant* to a task if and only if its relevancy to the task is 0. Examples of sensors irrelevant to a task include sensors whose sensing region have no overlap with the desired service region of a task, and sensors that cannot provide the type of information the task requires, such as a sensor providing temperature readings to an air pressure related task. On the other hand, for the coverage requirement of a task, we say a sensor *covers* a task if and only if the computed relevancy is 1. By definition, a sensor covers a task if and only if it can individually satisfy the desired QoI of the task. In a CPS sensory environment, the retrieved information from a single relevant sensor may not satisfy all QoI requirements of a task, thus resulting in a relevancy value that lies between 0 an 1. Therefore, to fully satisfy a task's QoI requirement, fusing information collected from multiple coordinating sensors will be needed.

6.4.1 Information Fusion

Some QoI requirements, like the coverage of a region, can be achieved by using a fusion algorithm (function) even if no individual sensor can meet the requirement. The authors in [169] propose to select a number of providers that cumulatively provide the most relevant information using an abstract, scalar-valued representation of QoI. While similar in principle, here we consider a more general way to accommodate a vector-valued QoI in information fusion. For ease of presentation, we use $g(\cdot)$ for the generic fusion function, and clearly, g should take a variant number of single-sensor "capabilities" and output an aggregated capability in all aspects. Note that there are a number of works on information fusion in WSNs (e.g., [170, 171]) that can be applied as a realization of $g(\cdot)$; further elaboration of $g(\cdot)$ is beyond the scope of this chapter. Denoting the capability of a subset $\mathcal{S}$ of sensors by $\underline{c}_{\mathcal{S}}$, we have:

$$\underline{c}_{\mathcal{S}} = g\Big(\{\underline{c}_n | n \in \mathcal{S}\}\Big). \tag{6.2}$$

Then, the relevancy of a subset of sensors to a task can be defined in the same way as that of a single sensor to a task based upon their aggregated sensing capability, i.e.,

$$r_{\mathcal{S},m} = f\Big(\underline{c}_{\mathcal{S}}, \underline{q}_m\Big), \quad \forall \mathcal{S} \subseteq \mathcal{N}, m \in \mathcal{M}. \tag{6.3}$$

6.4.2 Critical Covering Set

We define the *critical covering set* (CCS) of a task as a set of sensors whose aggregated sensor-to-task relevancy always achieves 1; and if the retrieved information from any sensor is lost, the aggregated relevancy will drop below 1. It is worth noting that there is finite probability that the sensors may not be able to cover the entire area of interest when randomly deployed. Furthermore, the desired QoI of certain tasks may be so demanding that even multiple collaborated sensors could not satisfy it. Therefore, it is possible that a task has no CCS. In this chapter, we focus on the scenario in which there is sufficient density of deployed sensors to always guarantee the existence of

CCSs for each task, with the realization that the system performance metric we defined in Section 6.5.2 also fits the scenario in which there exists no CCS for certain tasks. For ease of presentation, let $\mathbb{S}_m, \forall m \in \mathcal{M}$, be the set of all CCSs for task m and $\underline{\mathbb{S}} = \{\mathbb{S}_m\}$ the collection of all these sets.

6.5 QoI-Aware Energy Management

As discussed earlier, in order to fully exploit the energy-efficiency in a CPS sensory environment without sacrificing the QoI delivered to a task, both (a) the irrelevant sensors, i.e., sensors that are not relevant to any future incoming tasks, and (b) the redundant sensors, i.e., sensors that are not critical to any tasks, are allowed to be switched to the sleeping mode (`OFF`). In this section, we propose a framework to control the duty-cycling of these sensors based upon the task model outlined in Section 6.3.

6.5.1 Duty-Cycling of Sensors

The duty-cycle of a sensor is defined as the fraction of time that the sensor is `ON`, i.e., $\Sigma_n(T)/T, \forall n \in \mathcal{N}$, where $\Sigma_n(T)$ is the aggregation of the `ON` times during the lifetime T. We express the aggregated `ON` time as a function of T to explicitly describe its dependency on the lifetime. However, this straightforward definition of duty-cycle does not directly reflect the energy spent while switching between the two modes. Therefore, we propose a *generalized duty cycle* to explicitly incorporate the extra energy penalty paid each time the sensors switch modes. Specifically, let $P_n^{\text{sw}}, P_n^{\text{on}}$ denote the amount of energy consumed when each time sensor switches modes and keeps awake, respectively, and $N_n(T)$ is the number of mode switchings sensor n makes. Then the generalized duty cycle η_n of sensor n is defined as

$$\eta_n = \frac{P_n^{\text{sw}}}{P_n^{\text{on}}} \cdot \frac{N_n(T)}{T} + \frac{\Sigma_n(T)}{T}, \quad \forall n \in \mathcal{N}. \tag{6.4}$$

The goal of energy management is to minimize the generalized duty cycle in a CPS sensory environment, without sacrificing the QoI levels attained. At the beginning of each frame, the CPS EMS informs the gateway on the decisions as to when to switch modes for sensors in the current frame. Let $\underline{A}(t) = \{a_n(t)\},\ 0 \le a_n(t) \le L,\ n \in \mathcal{N}$ denote the mode switching times of sensors in the frame following the decision at time t. When $a_n(t) < L$, the n-th sensor will switch mode at time $t + a_n(t)$ and, when $a_n(t) = L$, it will not switch mode in this frame. Clearly, the cardinality of the decision space of $\underline{A}(t)$ is NL.

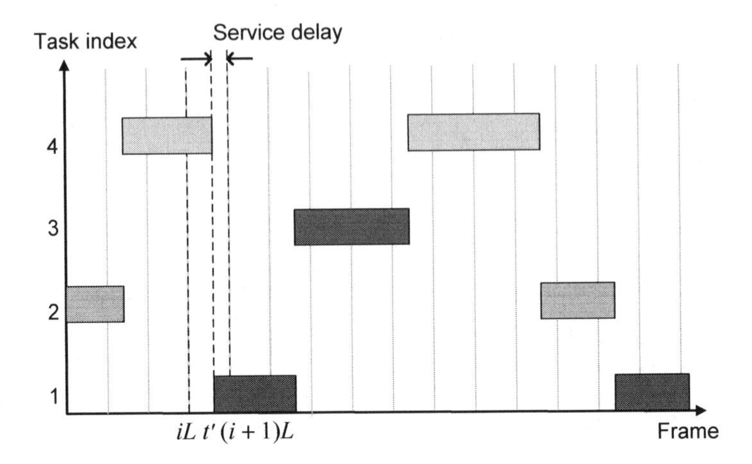

Figure 6.2: An illustrative example of service delay, where Tasks 1, 2, and 4 have two instances and Task 3 has only one instance.

6.5.2 *Delay Model for Tasks*

In practice, when the gateway sends the wake-up signal to the corresponding sensors they may not receive it immediately and be activated exactly at the scheduled time, mainly caused by the signal transmitting and processing latency. Moreover, even if the latency is so small as to be neglected, it is likely that no active CCS of a task exists when task instances start. Therefore, the task may have to wait for the next frame when the EMS informs the gateway to wake up a CCS for its service, as shown in Figure 6.2. Toward this end, we introduce the delay model for all tasks as follows.

We denote $d_{m,i}$ as the "attained" service delay of task m for its i-th instance. Note that this service delay may be tolerable, depending on the type of requested application (e.g., a few seconds' delay for reporting the water quality levels are highly likely tolerable). We denote D_m as the maximum tolerable delay for any instance of task m, defined as the fraction of time compared to the lifetime of the task instance:

$$D_m = \tau_m l_{m,i}, \ \forall m \in \mathcal{M}, \ i \in \mathbb{N}^+, \tag{6.5}$$

where $l_{m,i}$ is the lifetime of task m's i-th instance, and $\tau_m \in [0, 1]$ is the *delay tolerance* of task m. Clearly, smaller τ_m represents a more stringent delay requirement. Then, let $\zeta_{m,i}$ indicate if the system fails to satisfy task m's i-th instance's delay requirement; we have:

$$\zeta_{m,i} \triangleq \begin{cases} 0, & \text{if } 0 \leq d_{m,i} \leq \tau_m l_{m,i} \\ 1, & \text{if } d_{m,i} > \tau_m l_{m,i}, \end{cases} \tag{6.6}$$

$\forall m \in \mathcal{M}, \ i \in \mathbb{N}^+$. Therefore, considering the overall service delay for a task, its average measured *delay failure* probability is defined as:

$$\zeta_m = \frac{1}{I}\sum_{i=1}^{I}\zeta_{m,i},\ \forall m \in \mathcal{M},\ I \in \mathbb{N}^+, \tag{6.7}$$

and if the "attained" delay failure probability is smaller than a threshold ξ_m, we call its delay requirement successfully satisfied:

$$\zeta_m \leq \xi_m,\ \forall m \in \mathcal{M}. \tag{6.8}$$

Hence, we have introduced the delay model for all tasks in a probabilistic manner with the task-specified parameters ξ_m (delay failure threshold) and τ_m (delay tolerance).

6.5.3 Problem Formulation

At the beginning of each frame, the EMS informs the gateway of decisions made on the energy consumption state of each sensor $n \in \mathcal{N}$, i.e., which set of sensors should be awakened for task service in the current frame, and which set of sensors are allowed to be turned OFF, given the historical task evolution and sensor activity information, which we denote by $\underline{\mathbf{H}}(t)$. For ease of presentation, we use Λ to denote a generic task evolution model without specifying the mathematical details. Therefore, a decision policy ν is defined as a mapping from $\underline{\mathbf{H}}(t)$ to $\underline{A}(t)$, given the known task model and the predetermined CCS information:

$$\underline{A}(t) = \nu(\underline{\mathbf{H}}(t)|\Lambda, \underline{\mathbf{S}}). \tag{6.9}$$

The goal of EMS algorithm is to find the optimal decision policy ν^* that optimizes the sensor duty-cycles under the delay failure threshold for tasks. We propose two performance metrics to describe the system performance, and then formulate two corresponding optimization problems.

6.5.3.1 Minimize the Maximum Duty Cycle

As a collection of N sensors comprise the CPS sensory environments, the optimization of one single sensor duty-cycle does not represent the overall optimum, and this model starts from the overall CPS lifetime perspective that aims at providing a degree of fairness among all sensors (or in other words, the usages of all sensors are relatively comparable). The optimization problem is:

$$\begin{aligned} \underset{\nu}{\textit{minimize:}} \quad & \eta_{\max} = \max_{n \in \mathcal{N}} \eta_n \\ \textit{subject to:} \quad & \zeta_m \leq \xi_m, \quad \forall m \in \mathcal{M}, \end{aligned} \tag{6.10}$$

where the constraint is that the average measured delay failure for task m should not be larger than the tolerable threshold ξ_m, $\forall m \in \mathcal{M}$.

6.5.3.2 *Minimize Weighted Average Duty Cycle*

As the task is changing from time to time, some sensors may be excessively depleted, which can greatly decrease the system lifetime. This model aims at minimizing the average duty cycle of the entire CPS sensory environment, while taking the energy consumption fairness into consideration. The optimization problem is:

$$\begin{aligned} &\underset{\nu}{\textit{minimize:}} \quad \overline{\eta} = \sum_{n \in \mathcal{N}} \beta_n \eta_n, \\ &\textit{subject to:} \quad \zeta_m \leq \xi_m, \quad \forall m \in \mathcal{M}, \end{aligned} \tag{6.11}$$

where $\{\beta_n\} \in [0, 1]$ are weight factors, and $\sum_{n \in \mathcal{N}} \beta_n = 1$. We explicitly define it as the normalized ratio between the remaining energy $\bar{E}_n(t)$ and total energy reserve E_n to achieve a degree of energy fairness among all sensors:

$$\kappa_n = \exp\left(- \frac{\bar{E}_n(t)}{E_n} \right), \quad \beta_n = \frac{\kappa_n}{\sum_{n=1}^{N} \kappa_n}, \quad \forall n \in \mathcal{N}. \tag{6.12}$$

The mapping function is chosen to characterize the effect of decreased β_n with the increase of $\bar{E}_n(t)$. Clearly, β_n is a non-increasing function that the sensors with less residual energy are assigned with higher weights, indicating the smaller probability to be utilized at the decision moment. Therefore, a certain level of energy consumption fairness can be achieved, and network lifetime, defined as the time when the first sensor depletes its energy, is prolonged.

6.5.4 A Greedy Algorithm

The optimization problems in (6.10) and (6.11) are generally NP-hard and their optimal solutions are difficult to find without an exhaustive search. In this work, we propose a greedy algorithm for optimization problem (6.11). The algorithm is greedy in that at any decision point, it chooses the action that leads to the least *marginal increment* in $\overline{\eta}$.

For ease of presentation, suppose the system starts at $t = 0$. Denote $t = iL$ as the beginning of the i-th frame, where $i \in \mathbb{N}$. Denote η_t^n as the *runtime* generalized duty cycle of sensor n up to time t. η_t^n can be updated recursively by

$$\begin{aligned} \eta_n(t) \quad = \quad & \frac{1}{t} \left[\frac{P_n^{\text{sw}}}{P_n^{\text{on}}} \Big(N_n(t) - N_n(t-L) \Big) \right. \\ & \left. + \Big(\Sigma_n(t) - \Sigma_n(t-L) \Big) + \eta_n(t-L) \cdot (t-L) \right], \end{aligned} \tag{6.13}$$

$t = iL, i \in \mathbb{N}^+$, with $\eta_n(0)$ defined to be zero. Note that $N_n(t) - N_n(t-L)$ and $\Sigma_n(t) - \Sigma_n(t-L)$ are the number of state switches and the aggregated `ON` time between time $t - L$ and time t for sensor n, respectively. We define the marginal increase in the normalized energy consumption of sensor n between time t and $t+L$ as:

$$\Delta_n(t) \triangleq \frac{P_n^{\text{sw}}}{P_n^{\text{on}}}\Big(N_n(t+L) - N_n(t)\Big) + \Big(\Sigma_n(t+L) - \Sigma_n(t)\Big), \tag{6.14}$$

and clearly, we have:

$$\eta_n(t) = \frac{1}{t}\Big(\Delta_n(t-L) + (t-L)\cdot\eta_n(t-L)\Big), \quad \forall n \in \mathcal{N}. \tag{6.15}$$

Further, define the weighted average marginal increase in the normalized energy consumption of all sensors between time t and $t+L$ as

$$\overline{\Delta}(t) \triangleq \sum_{n\in\mathcal{N}} \beta_n(t)\Delta_n(t). \tag{6.16}$$

At the i-th decision point, EMS needs to predict the task activities in the current frame and prepare the sensors accordingly. Rather than a global algorithm that minimizes $\overline{\eta}$ throughout the lifetime of the CPS sensory environment, we specify an algorithm that minimizes $\overline{\Delta}(t)$ at each decision point while satisfactorily guaranteeing the service delay requirement, i.e., the delay failure probability.

Specifically, define $\Phi_{m,i}(t)$ as the probability that the i-th instance of task m starts exactly at time t, and $\Psi_m(t)$ as the "preparation" probability that at least one CCS of task m exists at time t. Then, we can rewrite the measured delay failure probability in Equation 6.7 as the sum of the probabilities when CCS of a task is not prepared under the condition of the task appearance:

$$\zeta_m = \sum_{t=0}^{T} \Phi_{m,i}(t)\Big(1 - \Psi_m(t)\Big), \quad \forall m \in \mathcal{M}. \tag{6.17}$$

T denotes the task lifetime. Note that $\Phi_{m,i}(t)$ is zero almost everywhere. To see this, $\Phi_{m,i}(t) = 0$ if: (a) t is not a task transition time, (b) either less than $i-1$ or greater than $i+1$ instances of task m have occurred. In other words, $\Phi_{m,i}(t)$ takes non-zero value only at the time of task transition and task m is expecting its i-th instance. Therefore, the above summation is easy to compute.

The task transition is modeled as a (discrete) semi-Markov process. A semi-Markov process is a stochastic process which moves from one state to another, with the successive states visited forming a Markov chain, and that the process stays in a given state a random length of time (holding time). The state space of a semi-Markov process is countable and the distribution function of the holding times may depend on the current state as well as on the one to be visited next [172]. When modeling the task evolution by a semi-Markov model, the tasks are treated as the states. The behavior of the tasks can be summarized in the following three aspects:

- There is at most one task in service at any time slot. And because of this, we call it the *exclusive task model*. In the main context, we only consider the existence of task instance, and we discuss the inclusion of "idle task" in Section 6.8;
- A new task starts immediately after a current task ends with certain "task transition" probability;
- The service time of a task is known at the time it starts.

We denote $\mathbf{P} = \{p_{k,m}\}$ as the task transition matrix, where $p_{k,m}$ is the transition probability from task k to task m. We also assume that $\mathbf{P}$ is known *a priori* to the gateway. In reality, $\mathbf{P}$ can be estimated based on the task evolution history by the EM algorithms[1] [173], which finds maximum likelihood or maximum a posteriori (MAP) estimates of parameters in statistical models in an iterative manner. The input of the EM algorithm is the observed data from task evolution history, consisting of the time of occurrence and departure of each task instance. More details will be given in Section 6.8.

Lemma 6.1
For any given task m, its delay requirement is satisfied, if the probability of any instance of it fails to provide the satisfactory service delay, τ_m, is upper-bounded by $\frac{\xi_m}{\pi_m}$, where ξ_m denotes the delay failure threshold, and π_m denotes the steady state probability of task m.

Proof 6.1 Define the steady state of transition matrix $\mathbf{P}$ of the Markov Chain that models the task evolution as $\boldsymbol{\pi}$. Due to the structure of the task model, all the possible states of the Markov Chain can be mutually visited, and thus, the Markov Chain is irreducible. Also, the Markov Chain is aperiodic since the probability of being at any state after N and $N+1$ transitions is greater than zero. Then, for an irreducible, aperiodic Markov Chain, there exists a steady state $\boldsymbol{\pi} = (\pi_1, \pi_2, \ldots, \pi_M)$ for all M tasks such that:

$$\boldsymbol{\pi}\mathbf{P} = \boldsymbol{\pi}, \text{ and } \sum_{m=1}^{M} \pi_m = 1. \tag{6.18}$$

This $\boldsymbol{\pi}$ can be found by solving the set of linear equations, and the result is:

$$\boldsymbol{\pi} = \mathbf{b}(\mathbf{B} - \mathbf{P} + \mathbf{I})^{-1}, \tag{6.19}$$

where $\mathbf{b}_{1\times M} = (1\ 1 \ldots\ 1)$, $\mathbf{B}_{M\times M}$ is an all-one matrix and $\mathbf{I}_{M\times M}$ is the identity matrix.

[1]The expectation maximization (EM) algorithms will be executed within the energy management server (EMS).

Then, given any task m, its average measured delay failure probability of all instances in Equation 6.7 can be rewritten as:

$$\begin{aligned}\zeta_m &= \lim_{I\to\infty}\frac{1}{I}\sum_{i=1}^{I}\zeta_{m,i} = \mathbb{E}\left[\zeta_{m,i}\right] \\ &= 1\cdot\Pr\{\zeta_{m,i}=1\}\pi_m + 0\cdot\Pr\{\zeta_{m,i}=0\}\pi_m \\ &= \Pr\{\zeta_{m,i}=1\}\pi_m,\ \forall m\in\mathcal{M}.\end{aligned} \tag{6.20}$$

Therefore, for satisfactory delay performance under parameter ξ_m, we rewrite $\zeta_m \leq \xi_m$ as:

$$\Pr\{\zeta_{m,i}=1\} \leq \frac{\xi_m}{\pi_m},\ \forall m\in\mathcal{M}, i\in\mathbb{N}^+. \tag{6.21}$$

In our model, the EMS knows exactly the time when the current task ends and the next task starts, whereas which specific task succeeds the current one is uncertain. If there is no transition between tasks in a frame, the system only needs to keep awake the CCS of the current task that leads to the least $\overline{\Delta}(t)$ and set the other sensors to sleep. Otherwise, if a task transition is bound to happen in a frame, the system has to wake up the corresponding sensors to make preparation for all possible succeeding tasks under their specific delay failure threshold. Suppose the current task will end during a frame and a new task will start at time t', where $t' \in [iL, (i+1)L)$. Our greedy algorithm at the i-th decision point is the solution to the following optimization problem

$$\begin{aligned}&\underset{\nu}{\textit{minimize:}} \quad \overline{\Delta}(iL) \\ &\textit{subject to:} \quad P_m(t')(1-\Psi_m(t')) \leq \xi_m, \quad \forall m\in\mathcal{M}.\end{aligned} \tag{6.22}$$

where $\overline{\Delta}(iL)$ is defined in Equations 6.14 and 6.16, and $P_m(t')$ is the transition probability from the current task to task m.

To solve the above optimization problem, the constraint can be rewritten as

$$\Psi_m(t') \geq 1 - \min\left\{1, \frac{\xi_m}{P_m(t')}\right\}, \tag{6.23}$$

which illustrates a way of computing the preparation probability given the task transition model and delay failure requirement. Essentially, Equation 6.23 specifies the minimum required probability of existence of CCSs for each task at the task transition time t'. Therefore, the EMS can determine whether to wake up a CCS for each task according to $\Psi_m(t')$, but the collective decisions for all tasks can be made either jointly or separately, since CCS may overlap for different tasks and a global optimum is achieved when the joint decision is made. However, this will induce further computational complexity, especially when M is large. Therefore, as another "degree of greediness," we let the EMS make decisions separately for each task. After the decision on making preparation for which tasks is made, the EMS chooses and schedules the wake-up times for a subset of sensors that can cover that selected group of tasks and induces the minimum increase in the marginal energy consumption $\overline{\Delta}(iL)$. In the $(i+1)$-th frame, which task follows the previous task is already known and all the irrelevant sensors prepared for the possible occurrence of other tasks can be sent to sleep.

It is worth noting that signal transmission and processing latency has impact on the control decision operation, and we shall investigate this in Section 6.7. As for now, we assume that when the decision is made at EMS and further informs the gateway, the latter is able to control all corresponding sensors immediately, i.e., the wake-up time of the scheduled sensors is the task transition time t'. If the selected sensors can cover the next arrival task rightly after the completion of the current task, at the start moment of the next frame $(i+1)L$, non-critical sensors will be shut down. However, if the predicted sensors are incapable of covering the actual coming task, at time $(i+1)L$, the gateway sends a new wake-up signal and activates the sensors from the CCS which induces the minimum increase in the marginal energy consumption $\overline{\Delta}((i+1)L)$. This is because the gateway already knows the identification of this task at time $(i+1)L$. It is easily obtained that, in the latter case, the service delay equals to $(i+1)L - t'$.

The algorithm can be summarized in the following steps:

1. At the beginning of each frame, shut down any sensor that is not critical to the current task, if there exist such sensors;

2. If no task transition is bound to happen in the current frame, keep the current status of each sensor until the next frame;

3. If a task transition is bound to happen in the current frame, for each task, compute the minimum required probability of existence of a CCS based on the delay failure threshold by Equation 6.23, and determine (by random tests) whether to make preparation for that task according to the derived probability. At the time of task transition, wake up a subset of sensors that critically covers all the tasks to be prepared for, yet induces the minimum increase in the marginal energy consumption.

The algorithm is greedy in three aspects:

1. The algorithm satisfies the delay failure threshold every time task transitions happen;

2. The algorithm minimizes the marginal increase in energy consumption at every decision point;

3. The algorithm makes a decision on whether to prepare for the possible occurrence of each task *separately*.

As discussed earlier, we can revise the greedy algorithm so that it makes a decision on whether to prepare for the possible occurrence of each task *jointly*.

Compared to the original one whose computational complexity increases linearly with M, the *revised greedy algorithm* increases exponentially with M and it is therefore much more time consuming. In order to show the potential improvement, we demonstrate the results for both greedy algorithms in Section 6.6.

6.6 Performance Evaluation

In this section, we show an example of our methodology, by assuming a monitoring CPS application, such as using (randomly deployed) pollutant-sensing sensors with certain sensing range to measure the water quality of certain locations in a reservoir supplying water to city dwellers. We present the system pertinent solutions, the environment settings, and show the results.

6.6.1 System Model and Simulation Setup

In our water quality monitoring system, each sensor with pollution level monitoring capability is randomly deployed and its spatial coverage follows a classic disk model. We assume that the sensory data within the sensing region is corrupted by noise during measurement and/or transmissions. Figure 6.3 shows an illustrative example of the reservoir plane graph for sensor and task deployment, where $N = 15$ sensors are deployed in a 600×600 square unit distance of area to monitor the water quality of $M = 4$ specific locations.

In this example, we consider measurement accuracy and service delay (both with multiple metrics) as two QoI requirements. For the former, we define its probabilistic model as:

$$\Pr\Big\{|Z_m(t) - z_m| \geq \delta_m\Big\} \leq \epsilon_m, \ \forall m \in \mathcal{M}, \tag{6.24}$$

where the random variable $Z_m(t)$ is the fused, sensor-retrieved information for task m at time t, and z_m is the actual but unknown information, i.e., the ground truth. Analogously to the desired QoI functions in [169], we define $\underline{q}_m$ as:

$$\underline{q}_m = \Big\{Y_m, (\delta_m, \epsilon_m)\Big\}, \ \forall m \in \mathcal{M}, \tag{6.25}$$

where Y_m and $\{\delta_m, \epsilon_m\}$ are the geographical location and accuracy requirement of task m, respectively. For service delay, tasks specify the required delay tolerance threshold τ_m and delay failure probability $\xi_m, \forall m \in \mathcal{M}$, as shown in Section 6.5.2.

On the other hand, the sensing capability of sensor n, i.e., $\underline{c}_n$, can be defined as:

$$\underline{c}_n = \Big\{(X_n, r_n), \gamma_n\Big\}, \ \forall n \in \mathcal{N}, \tag{6.26}$$

where X_n is the location of the sensor and r_n is its sensing radius. We model

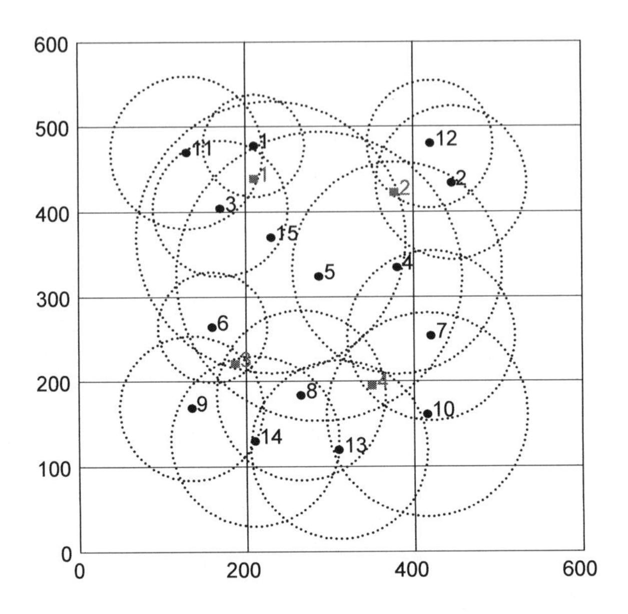

Figure 6.3: An illustrative example of the reservoir plane graph to monitor the water quality of four locations (as square), where 15 randomly deployed sensors with certain sensing range are shown (as dots).

the measurement noise as additive white Gaussian noise (AWGN) with variance γ_n for sensor n. A sensor-to-task relevancy function for this model is

$$\begin{aligned} f(\underline{c}_n, \underline{q}_m) &= f(X_n, r_n, \gamma_n, Y_m, \delta_m, \epsilon_m) \\ &= \mathbf{1}\{\text{dist}(X_n, Y_m) \leq r_n\} \cdot \min\left\{\frac{\epsilon_m}{\Pr\{|W_n(t) - \omega L| \geq \delta_m\}}, 1\right\} \\ &= \mathbf{1}\{\text{dist}(X_n, Y_m) \leq r_n\} \cdot \min\left\{\frac{2\epsilon_m}{Q(\frac{\delta_m}{\sqrt{\gamma_n}})}, 1\right\}, \end{aligned} \tag{6.27}$$

$\forall n \in \mathcal{N}, m \in \mathcal{M}$, where $\mathbf{1}\{\textit{statement}\}$ is the indicator function that takes value 1 if the *statement* is true and 0 otherwise, $\text{dist}(X_n, Y_m)$ is the Euclidean distance between two points, the random variable $W_n(t)$ is the information retrieved from sensor n at time t, and $Q(\cdot)$ is the tail probability of the standard normal distribution.

If task m is serviced solely by sensor n, then $Z_m(t) = W_n(t)$; otherwise, if it is serviced by a subset $\mathcal{S}$ of sensors, then $Z_m(t) = W_{\mathcal{S}}(t)$, where $W_{\mathcal{S}}(t)$ is the fused information from a subset $\mathcal{S}$ of sensors. One possible information fusion algorithm of relevant sensors in this case can be:

$$W_{\mathcal{S}}(t) = \arg\min_{w} \frac{1}{|S|} \sum_{n \in \mathcal{S}} \frac{1}{\gamma_n} \left| W_n(t) - w \right|^2 = \frac{\sum_{n \in \mathcal{S}} \frac{W_n(t)}{\gamma_n}}{\sum_{n \in \mathcal{S}} \frac{1}{\gamma_n}}. \tag{6.28}$$

The right-hand side of Equation 6.28 is a specific example of the fusion function $g(\cdot)$ we defined in Section 6.4.1. Specificially, if all γ_n are equal and $W_n(t) \sim \mathsf{N}(1, \gamma)$, the fused information of a group of K relevant sensors is the average of the individual ones and $W_t^S \sim \mathsf{N}(1, \gamma/K)$, where $\mathsf{N}(\mu, \sigma^2)$ is a Gaussian distribution with mean μ and variance σ^2. Based on the above fusion algorithm, CCSs of every task can be computed during the sensor deployment stage, and used in the online duty-cycling control.

Our numerical result is based on the water quality monitoring system discussed above and is achieved in MATLAB®. The capabilities (exclusive of sensing radius which is illustrated in Figure 6.3) of all sensors are: $\gamma_n = 3$, $\forall n \in \mathcal{N}$. Moreover, $P_n^{\text{sw}} = 5$ and $P_n^{\text{on}} = 1, \forall n \in \mathcal{N}$ and the initial energy reserve of each sensor is set as 20,000 units. We assume that with the predetermined working power, each sensor is able to fully cover its sensing area. For all tasks in $\mathcal{M}$, the desired QoI satisfies: $\epsilon_m = 0.1$, $\delta_m = 1$, $\tau_m = 0$ (i.e., delay-sensitive applications). Assume that the service time of each task follows identical exponential distribution with average duration $1/\mu = 50$ time slots and minimum duration $l_{\min} = 25$ time slots (both are sufficiently longer than the frame size L), thus the arrival of all tasks is a Poisson process. A total of 1,000 task instances are simulated. The task transition matrix is given by:

$$\mathbf{P} = \begin{pmatrix} 0 & 1/10 & 2/5 & 1/2 \\ 1/5 & 0 & 3/5 & 1/5 \\ 1/3 & 1/3 & 0 & 1/3 \\ 4/5 & 1/10 & 1/10 & 0 \end{pmatrix}. \tag{6.29}$$

The sensor-to-task relevancy and CCSs can therefore be computed at offline, and there are 10 candidate sets for Tasks 1, 2, and 4, and 20 CCSs for Task 3. Meanwhile, each CCS consists of three sensors.

We consider the optimization problem in Equation 6.11 with the energy-aware weight factor β_n in Equation 6.12, and the solution is obtained by using the proposed greedy algorithms outlined in Section 6.5.4.

6.6.2 Simulation Results

In Figure 6.4, we arbitrarily pick up five sensors and plot the evolving trend of their duty cycles over time. We set up the system parameters as stated above with $\xi_m = 0.04$, $\forall m \in \mathcal{M}$ and $L = 20$ time slots. It can be seen from the figure that after a few rounds of fluctuations at the very beginning when the sensors are trading-off their energy consumption with the provided QoI to tasks, the duty cycle of each sensor converges soon afterwards. This is because our proposed greedy algorithm successfully selects the best set of sensors for service under the stochastic, but Markovian task transitions, and the weight factors accurately capture the energy consumption state of all sensors and guarantee a degree of fairness among them.

Next, we show the impact of two system parameters, the frame size L and sensor mode switching power P_n^{sw}, on the average duty cycle of all deployed

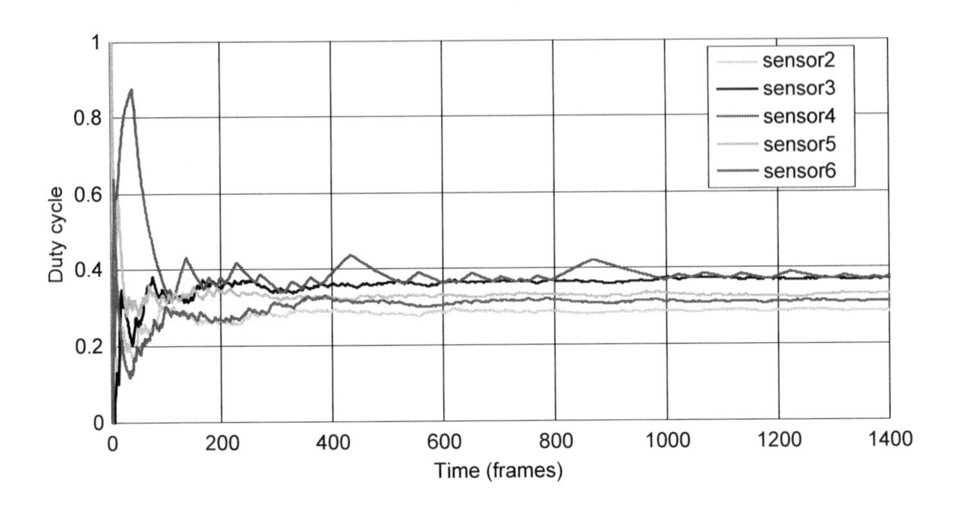

Figure 6.4: Sensor duty cycle *vs.* time

sensors as shown in Figure 6.5(a) and Figure 6.5(b), respectively. The delay failure thresholds ξ_m of all tasks are equally chosen while varying between 0 and 0.1. We observe that for fixed L and P_n^{sw}, the average measured duty cycle linearly decreases with the increase of required delay failure threshold, as higher ξ_m relaxes the service delay requirement provided to all tasks by allowing a certain number of task instances to fail, and in turn the sensors spend more time in the sleeping mode. For fixed ξ_m, the duty cycle increases with L and P_n^{sw}. Larger L represents the less frequency system control decisions and thus in order to provide satisfactory services to the next task, the system tends to wake up more sensors than necessary. These unnecessary sensors will stay awake until the next decision point when they can be turned `OFF` by the EMS. Clearly, the wasted `ON` times of sensors increase linearly with the frame length. Furthermore, larger P_n^{sw} indicates the less reluctant control behavior (or higher penalty) when switching the mode. Thus, the system decisions favor those sensors who have been in the `ON` state, and let them continue servicing other tasks that may not eventually appear. Therefore, the energy consumption of all sensors is not optimally allocated, resulting in larger average duty cycle and its variance.

Similar trends have been found when simulating the impact of task accuracy requirement ϵ_m, as shown in Figure 6.5(c). With the increase of ϵ_m, i.e., less stringent QoI requirement that allows more measurement errors, the CCS of a task may involve fewer sensors for service, and in turn reduce the average duty cycle.

Then, we investigate the impact of system parameter L on the measured delay failure probability given different delay failure thresholds, i.e., to judge if the required delay parameters are successfully guaranteed by the greedy

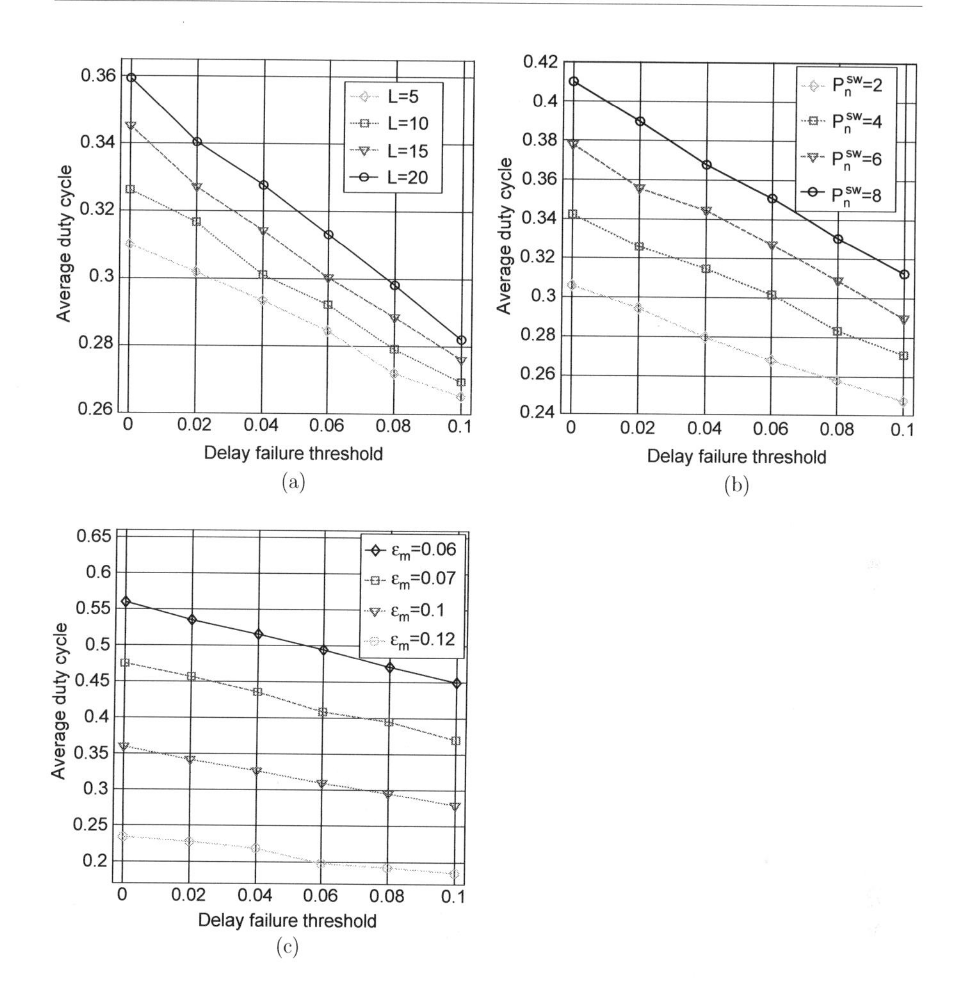

Figure 6.5: Average duty cycle *vs.* delay failure threshold, by changing (a) frame size $L = \{5, 10, 15, 20\}$, (b) switching power $P_n^{\mathrm{sw}} = \{2, 4, 6, 8\}$, and (c) task accuracy requirement $\epsilon_m = \{0.06, 0.07, 0.10, 0.12\}$.

algorithm. Table 6.3 shows the result, where the measured delay failure is satisfactorily less than the delay failure threshold and apparently, it increases with the threshold. However, it is interesting to observe that under the same delay failure threshold, the difference between measured results of different frame sizes L is insignificant. This can be explained by our setting of $\tau_m = 0$, $\forall m \in \mathcal{M}$, i.e., we only consider the delay-sensitive applications, and thus as long as the prepared sensors are incapable of servicing the coming task at the task transition time, the delay failure event is counted, irrespective of how big the frame size L is (or when the next decision time will be, even if the sensors are well-prepared then). We shall investigate the impact of signal transmission

Table 6.3: Average Measured Delay Failure w.r.t. Different Frame Sizes and Delay Failure Thresholds

	$L = 5$	$L = 10$	$L = 15$	$L = 20$
$\xi_m = 0$	0	0	0	0
$\xi_m = 0.02$	0.0150	0.0133	0.0150	0.0137
$\xi_m = 0.04$	0.0338	0.0298	0.0295	0.0312
$\xi_m = 0.06$	0.0445	0.0428	0.0428	0.0468
$\xi_m = 0.08$	0.0655	0.0583	0.0637	0.0645
$\xi_m = 0.1$	0.0798	0.0755	0.0785	0.0795

and processing latency and frame size L on delay-tolerable applications in the next section.

Figure 6.6 illustrates the energy depletion process, for $\xi_m = 0.04$, $L = 20$, with other parameters being set up as the former setting. Besides the proposed greedy algorithm and its revised version (i.e., jointly considering the CCSs of all tasks rather than treating them separately), we also show the result of the optimal solution where the EMS knows exactly which task succeeds the current one. The slope of a curve in the figure represents the energy depletion rate. We observe that the revised greedy algorithm only achieves a slightly better performance than the basic greedy algorithm, at the expense of more computational complexity. The gap between the greedy algorithm and the genie-aided optimal solution can also be identified, due to the potential error in estimating the future arrived task. Furthermore, we plot the energy depletion process for two extremes: the least (sensor 12) and most used (sensor 5) sensor in the proposed greedy algorithm. As sensor 12 is located at the border area with limited sensing range, it can only cover task 2, which is also being covered by many other sensors like 2, 4, 5, 15, and thus being least frequently used. Meanwhile, sensor 5 is located at the center with a relatively larger sensing radius allowing it to service all four tasks, thus being used mostly. Nevertheless, the energy-aware weight factor in Equation 6.12 that explicitly takes the residual energy of a sensor into consideration helps to lower the variance between these two extremes so that a certain degree of energy consumption fairness is achieved, as shown in the following.

We investigate the energy consumption fairness, quantified by the Jain's fairness index[2] under the proposed greedy algorithm as shown in Figure 6.7. We compare the proposed energy-aware weight assignment approach with the equal weight assignment, i.e., $\beta_n = 1/N, \forall n \in \mathcal{N}$. Clearly, for a fixed number of sensors N, the Jain's index under energy-aware approach is higher than the one under equal assignment. Furthermore, when more nodes are deployed

[2]It is defined by $(\sum(E_n - \bar{E}_n))^2/(N\sum(E_n - \bar{E}_n)^2), \forall n \in \mathcal{N}$. The result ranges from $\frac{1}{N}$ (worst case) to 1 (best case). The larger the index is, the better fairness we can achieve.

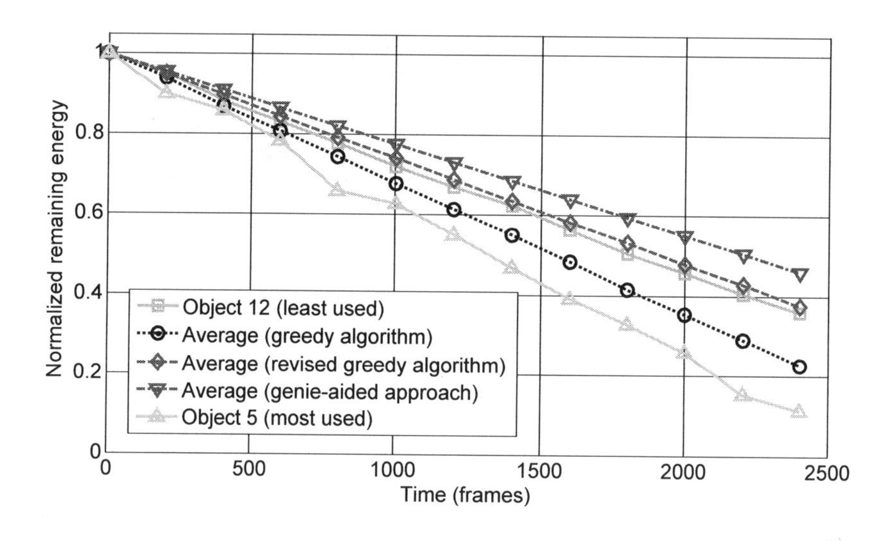

Figure 6.6: Normalized remaining energy *vs.* time.

in a fixed geographic area, the increased node density helps to achieve better fairness among them since any single task would potentially be serviced by more CCSs. This trend does not hold for the equal weight setting, since the diversity gain cannot be utilized by assigning the same weights to all sensors irrespective of their different amounts of residual energy, and in turn the fairness level decreases with the number of sensors.

6.7 Modeling the Signal Transmission and Processing Latency

In practice, because of the signal transmission and processing latency by MAC protocols and routing algorithms, the selected CCS members cannot be awakened immediately after the control decision is made. Furthermore, for many applications they do not require immediate task service, but allowing certain service delay after the specified start time. Toward this end, in this section, we improve our system model by explicitly considering the signal transmission and processing latency, modeled by certain amount of wake-up delays after the control decision is made when all CCS members have been successfully informed. In other words, this model considers the longest signal propagation delay from EMS to a CCS member, denoted as a period of ωL, $\omega \in \mathbb{Z}$. We also assume that once the wake-up signal has been sent, it cannot be revoked. In this section, we provide a thorough theoretical analysis on the new system model, coupled with experimental results of the impact of this signaling latency on average measured delay failure probability.

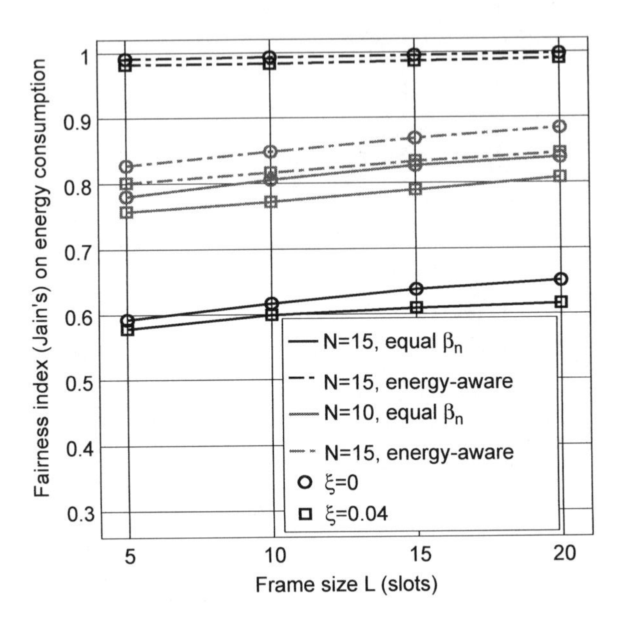

Figure 6.7: Fairness index (Jain's) on energy consumption among all sensors.

6.7.1 Model Description and Problem Formulation

Without loss of generality, we rename the appearance sequence of all task instances sequentially by index $k = 1, 2, \ldots$, as shown in Figure 6.8. The figure also illustrates a localized view on a specific time period, where the k-th instance of all tasks arrives at time $t' \in (iL, (i+1)L)$, $i \in \mathbb{N}^+$, with its lifetime l_k. Further let x denote the interval between iL and t' as a random variable; then we have $t' = \sum_{i=1}^{k-1} l_i = iL + x$.

After receiving the control decision from EMS, the gateway sends a wake-up signal to the corresponding sensors at decision time iL. If at least one CCS of instance k exists at time t', the CCS members will be awakened at time $(i+\omega)L$, i.e., after this control decision successfully propagates to all sensors in ωL frames. Then, the service delay is computed as:

$$(i+\omega)L - t' = (i+\omega)L - (iL + x) = \omega L - x. \tag{6.30}$$

However, if no CCS of instance k exists at time t', by knowing exactly which task will appear at the next decision time $(i+1)L$, the gateway then sends a new wake-up signal which eventually takes effect at time $(i+1+\omega)L$, i.e., after considering the signal transmission and processing latency $(1+\omega)L$ from the current time. In this case, the service delay is:

$$(i+1+\omega)L - t' = (\omega+1)L - x. \tag{6.31}$$

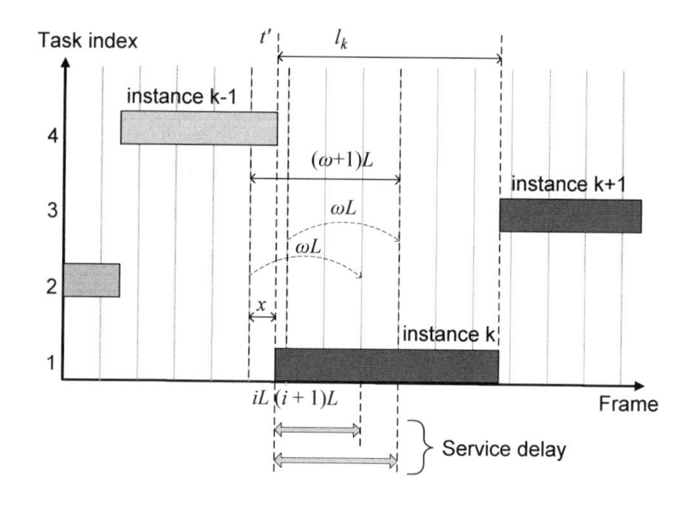

Figure 6.8: An illustrative example of service delay with signaling latency.

Note that it is necessary to ensure $(\omega + 1)L - x < l_k$, $\forall k \in \mathbb{N}^+$, since the task instance k needs to be handled before its termination. Consequently, $\omega < \lfloor l_{\min}/L \rfloor - 1$.

We are able to compute the probability that the service delay incurred at any instance is larger than the specified delay tolerance $\tau_m l_k$, conditioned when a CCS is well-prepared for service and otherwise, as:

$$P_m(t')\Psi_m(t')\Pr\{\omega L - x > \tau_m l_k\} \\ +P_m(t')\left(1 - \Psi_m(t')\right)\Pr\{(\omega + 1)L - x > \tau_m l_k\} \leq \xi_m, \quad (6.32)$$

$\forall m \in \mathcal{M}$. It is worth noting that Equation 6.32 exactly characterizes the constraint of our optimization problem in Equation 6.22, and more importantly a way to further theoretically derive the left-hand side of Lemma 6.1.

Theorem 6.1

Given the service time distribution in the previous assumptions, $\Psi_m(t')$ in (6.32) can be derived as:

$$\Psi_m(t') \geq \frac{F(\omega + 1) - \frac{\xi_m}{P_m(t')}}{F(\omega + 1) - F(\omega)}, \quad (6.33)$$

where

$$F(\omega) = 1 - \frac{\tau_m}{\mu L}\exp\left(-\frac{\omega\mu L}{\tau_m} + \mu l_{min}\right)\left(\exp\left(\frac{\mu L}{\tau_m}\right) - 1\right). \quad (6.34)$$

Proof 6.2 As the service time of each task follows identical exponential distribution, the total number of instance occurrences has a Poisson distribution over $(0, t]$, and the occurrences are distributed uniformly on any interval of time. Therefore, the random variable x shown in Figure 6.8 follows a uniform distribution in $(0, L)$. As l_k is exponentially distributed with average $1/\mu$ and lower-bound $l_{\min}$, its probability density function is given by:

$$f_l(l) = \begin{cases} \mu \exp(-\mu l + \mu l_{\min}), & l > l_{\min}, \\ 0, & \text{others.} \end{cases} \tag{6.35}$$

From Equation 6.32, we have:

$$\Psi_m(t') \geq \frac{\Pr\{(\omega+1)L - x > \tau_m l_k\} - \frac{\xi_m}{P_m(t')}}{\Pr\{(\omega+1)L - x > \tau_m l_k\} - \Pr\{\omega L - x > \tau_m l_k\}}. \tag{6.36}$$

Let $F(w)$ denote the probability that the incurred service delay is larger than the specified tolerance, or $F(w) \triangleq \Pr\{\omega L - x > \tau_m l_k\}$, then:

$$\begin{aligned} F(w) &= \Pr\{l_k < \frac{\omega L - x}{\tau_m}\} \\ &= \int_0^L \frac{1}{L} \int_{l_{\min}}^{\frac{\omega L - x}{\tau_m}} \mu \exp(-\mu l_k + \mu l_{\min}) dl_k \, dx \\ &= 1 - \frac{\tau_m}{\mu L} \exp\left(-\frac{\omega \mu L}{\tau_m} + \mu l_{\min}\right)\left(\exp\left(\frac{\mu L}{\tau_m}\right) - 1\right). \end{aligned} \tag{6.37}$$

Hence, replacing Equation 6.37 back to Equation 6.36, we obtain the closed-form expression for $\Psi_m(t')$ in Equation 6.33.

We next discuss the feasibility issues of Theorem 6.1. As $F(\omega)$ monotonically increases with ω, we have $F(\omega+1) > F(\omega)$ always holds. Since $\Psi_m(t')$ denotes the *preparation probability* that at least one CCS of task m exists at time t', by definition it is a scaler between 0 and 1. Therefore, it requires:

$$F(\omega) \leq \frac{\xi_m}{P_m(t')} \leq F(\omega+1). \tag{6.38}$$

However, since $\xi_m \in [0, 1)$ is the maximum allowed delay failure probability, it is specified by applications, and does not have relations with the transition probability $P_m(t')$ from the current task to task m. Therefore, Equation 6.38 may not always hold. Recall that $F(\omega)$ represents the probability that the incurred service delay is larger than the specified tolerance, or $F(\omega) \triangleq \Pr\{\omega L - x > \tau_m l_k\}$. Then, when $\frac{\xi_m}{P_m(t')} > F(\omega+1)$, we set $\Psi_m(t') = 0$. This is because if the maximum allowed delay failure is relatively very high or the task arrival probability is low enough, there is no need to prepare sensors for it. On the contrary, when $\frac{\xi_m}{P_m(t')} < F(\omega)$, we set $\Psi_m(t') = 1$ as a constant, indicating that if the maximum allowed delay failure is very low or a task

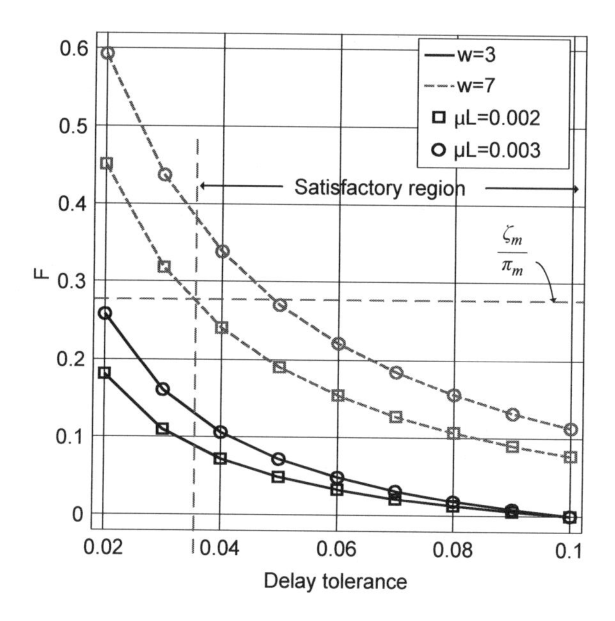

Figure 6.9: F *vs.* delay tolerance threshold, varying parameters $\omega = \{3, 7\}$ and $\mu L = \{0.002, 0.003\}$.

instance is very likely to come at time t', it is necessary to prepare sensors for service.

In summary, replaced by the new constraint in Equation 6.33 under the realistic delay model, our objective function in Equation 6.22 and proposed greedy algorithms can provide a sub-optimal solution. All other steps in Section 6.5.4 apply.

6.7.2 Satisfactory Region of Delay Tolerance

The value of F denotes the theoretically derived probability that the service delay exceeds the maximum tolerable threshold. Figure 6.9 shows the value of F with respect to (w.r.t.) different specified delay tolerance values, by varying ω and μL. Consistent with previous analysis, F monotonically increases with ω, since more severe signaling latency would result in higher delay outage probability. It also can be seen from the figure that higher delay tolerance τ_m leads to lower delay probability, and this probability increases with μL that characterizes the ratio of frame size and average duration of task instance. Higher μL (i.e., larger frame size or shorter instance duration) will relax the delay constraint imposed by tasks, and thus lower delay outage is expected.

As analyzed in Lemma 6.1, given task m, in order to successfully achieve its delay requirement, the probability of delay occurrence at any of its instances should be upper-bounded by ξ_m/π_m. Based on our analysis of signaling latency in Equation 6.32, we rewrite the steady state form of Lemma 6.1 as:

$$\Psi_m F(\omega) + (1 - \Psi_m) F(\omega + 1) \leq \frac{\xi_m}{\pi_m}, \tag{6.39}$$

where Ψ_m denotes the corresponding preparation probability under the steady state of task transitions. Since $F(\omega) < F(\omega + 1)$, we relax the constraint in Equation 6.39 by

$$F(\omega) < \frac{\xi_m}{\pi_m}, \quad \forall m \in \mathcal{M}. \tag{6.40}$$

According to the derivation of F as a non-increasing function of the delay tolerance τ_m, it is interesting to observe that τ_m cannot be arbitrarily chosen, but tightly coupled with system parameters $L, \omega, \mathbf{P}_m$ and task parameters $\mu, l_{\min}$. In Figure 6.9, we visualize the condition (Equation 6.40) that eventually crosses all curves of different parameters. We call the region of τ_m satisfying the condition (Equation 6.40) as its *satisfactory region*. Therefore, given those parameters, the system has its own feasible working range, beyond which higher system settings (like L) should be configured. Deriving this lower-bounded region requires solving the transcendental equation and thus, numerical solutions are expected like the Newton's method.

6.7.3 Results

We still use the same task transition matrix $\mathbf{P}$ as shown in Section 6.6.1, and set $\xi_m = 0.1, \forall m \in \mathcal{M}$. A total of 1,000 task instances are simulated, with average duration of $1/\mu = 2000$ and minimum duration $l_{\min} = 25$ time slots. Other parameters are the same as the basic setting in Section 6.6.1.

According to the proof of Lemma 6.1, the steady state $\boldsymbol{\pi}$ is obtained as $(0.33, 0.14, 0.25, 0.28)$. Each element of the steady state denotes the stationary probability of a specific task. Figure 6.10 shows the simulation result on the average measured delay failure among all tasks w.r.t. different delay tolerance thresholds. It can be observed that $\omega = 3$ can successfully guarantee that the required $\xi_m = 0.1$ for all $\tau_m \in [0.02, 0.1]$; however only part of entire τ_m values can satisfy the same requirement when $\omega = 7$, consistent with our analysis in Section 6.7.2. Furthermore, smaller parameter μL decreases the probability of delay failure occurrences either through more frequent control decisions (smaller L) or servicing longer instance duration (larger $1/\mu$); equivalently wider satisfactory region for delay tolerance given a ξ and $\mathbf{P}$.

6.8 Implementation Guidelines

We have made a few simplifications and assumptions in Section 6.3 to ease the analysis, some of which may potentially generate new implementation guidelines in practice.

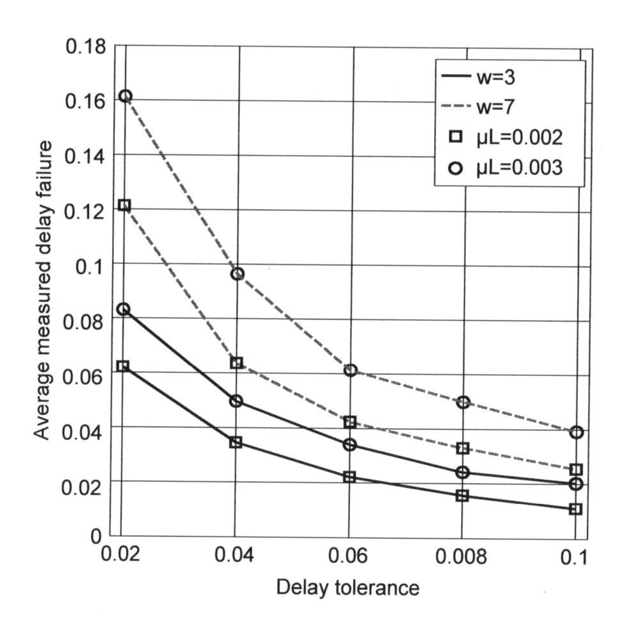

Figure 6.10: Average measured delay failure probability *vs.* delay tolerance threshold.

First, we assume that the CCS are known *a priori* to the EMS, which is usually realized during the deployment stage where the application owner deploys a certain number of devices to the specified network domain with known geographical locations. Then, the offline computation can be performed given the desired task location and requirements.

Second, the computation of task transition matrix requires advanced algorithms like EM algorithm [173]. It uses an iterative procedure to compute the maximum likelihood estimation of a set of parameters in a given distribution (from empirical analysis). To apply it in our framework, EM needs the observed data from task evolution history, including the start and end time of each task instance, from which we can derive the transition times between tasks. Then, the EM algorithm approximates the parameters of the given distribution, as well as its expected value. Note that these expected values exactly represent the average number of transitions between each pair of two tasks, whose normalized values are a "noisy" version of the hidden, true value of all entries of the transition matrix **P**.

Finally, in practical scenarios, tasks are described in human-friendly formats, e.g., XML/HTML, and thus a higher layer of format interpretation is used to translate the encoded scripts into the required QoI attributes, which can then be easily incorporated into our framework. Also, as a middleware bridging a variety of different applications and underlying networks, although tasks can be submitted randomly by each user, we assume that the aggregated behavior as the input to our platform exhibits some degree of determined characteristics, e.g., the total number of tasks, and their associated task QoI requirements. Moreover, there may not always be a ready task after the current task ends. Therefore, we can incorporate an "idle state" into the task transition model. The transition probability between idle state and other tasks can also be estimated by historical observations on task evolutions over time. When the idle task arrives, at the next decision point, we simply shutdown all prepared sensors and keep this state until the arrival of the next task instance. Therefore, all previous analysis and proposed algorithms still apply.

6.9 Summary

In this chapter, a system-level efficient energy management framework is proposed to provide a satisfactory QoI experience in CPS sensory environments. Contrary to past efforts, our proposal is transparent and compatible with lower protocols in use, and preserves energy-efficiency in the long run without sacrificing any attained QoI levels. Specifically, we introduced the new concept of QoI-aware "sensor-to-task relevancy" to explicitly consider the sensing capabilities offered by a sensor to the CPS sensory environments, and QoI requirements required by a task. Then, we proposed a novel concept of the "critical covering set" of any given task in selecting the sensors to service a task over time. Next, an energy management decision is made dynamically at runtime, as the optimum for long-term traffic statistics under the constraint of the service delay. An extensive case study based on utilizing the sensor networks to perform water level monitoring is given to demonstrate the ideas and algorithms proposed in this chapter, and a simulation is made to show the performance of the proposed algorithms. To make our energy management framework more applicable and practical in realistic scenarios, we further considered the signal transmission and processing latency into our system model, and both theoretically and experimentally showed its impact on average measured delay probability. Finally, based on our system model assumptions, we brought forward some implementation guidelines in practice and discussed the applicability of our proposal.

ENABLING TECHNOLOGIES FOR CPS

Chapter 7

Networking Technologies for CPS

Chi Harold Liu

Beijing Institute of Technology, China

Zhengguo Sheng

University of Sussex, U.K.

CONTENTS

The preceding chapters discuss the architecture of CPS. First we discuss the CPS architecture in general, and then separate it into different sub-sections, including the mobile sensing devices/platforms, naming, addressing and profile services, device search and selection, device management, and energy management. In the following chapters, we will consider the enabling technologies of CPS. First, we start with the fundamental networking technologies in this chapter.

	Technical Summary	Radio Band	Applications	Manufacturers
433MHz enabled proprietary solutions	Proprietary solutions by using one of the most commonly used ISM (industrial, scientific, and medical) radio bands in China.	433MHz	Home security (with China mobile), environment monitoring, etc.	Homewell Beelinker
ZigBee	A well-defined protocol stack for WSN with features of self-deployment, low complexity, low data rate, and low cost, etc.	780MHz, 2.4GHz	Smart Energy, Home Automation, Building Automation. Health care, Remote Control, Retail Services, etc.	Vinnotech, Smeshlink, Starvalley
RFID	A fast developing radio technology used to transfer data from an electronic tag, which includes identification, information collection, etc.	125KHz, 13.56MHz, 433MHz,	Logistic, E-car license, one pass card	Fudan microelec. Huahong, Vision electronics etc.
Bluetooth	Bluetooth low energy technology is a global standard, which enables devices with coin cell batteries to be wirelessly connected to standard Bluetooth enabled devices and services.	2.4GHz	Remote access, Indoor positioning (HAIP)	Nokia Research Beijing

Figure 7.1: Radio technologies and applications mapping.

7.1 Sensing Networks

It is worth noting that China has still little developed sensor technologies, especially in the manufacturing of high-precision sensor chips. With around 20k kinds of sensor chips around the world, China can only manufacture about 300 of them. Therefore, in this section, we are not going to focus on sensor chips, but on radio technologies and networking abilities.

In China, SMEs (small- and medium-sized enterprises) play a very important role in sensing networks in terms of technologies and products. Figure. 7.1 shows a list of radio technologies being used in M2M applications in China.

Figure. 7.2 shows the authorized ISM band in China. It is worth noting that most popular radio technologies for M2M applications are based on ISM.

7.1.1 433MHz Proprietary Solutions

433MHz is one of the most commonly used radio bands in industrial-grade M2M modules in China. It provides the wireless remote transmission capacity for various M2M terminals. The application area covers electric power, oil fields, coal mines, meteorology, environmental protection, water conservancy, heat power, gas, telecommunications, postal services, banking, transportation, petrochemicals, etc.

7.1.2 ZigBee

The Chinese manufacturers are on a relatively a small scale and in the beginning stage of M2M industry development. Major players currently focus on the development of mesh connection and proprietary protocol stacks. With the

ISM Band	Frequency Range	Center Frequency
1	6.765 MHz - 6.795 MHz	6.780 MHz
2	13.553 MHz - 13.567 MHz	13.560 MHz
3	26.957 MHz - 27.283 MHz	27.120 MHz
4	40.660 MHz - 40.700 MHz	40.680 MHz
5	433.050 MHz - 434.790 MHz	433.920 MHz
6	2.400 GHz - 2.500 GHz	2.450 GHz
7	5.725 GHz - 5.875 GHz	5.800 GHz
8	24.000 GHz - 24.250 GHz	24.125 GHz
9	61.000 GHz - 61.500 GHz	61.250 GHz
10	122.000 GHz - 123.000 GHz	122.500 GHz
11	244.000 GHz - 246.000 GHz	245.000 GHz

Figure 7.2: ISM (industrial, scientific, and medical) band in China.

development of ZigBee technology, especially the development of application profiles, more Chinese companies tend to join the camp of ZigBee.

7.1.3 RFID

China is a follower in RFID industry. However, the future demand of RFID is huge. The RFID driven M2M applications could be the most dominant and popular in China. It is widely used in logistics and intelligent transportation. Some key RFID manufacturers are listed in Figure 7.3.

It is worth noting that China has a well-developed RFID value chain. There are also related R&D and testing institutes, such as the Institute of Automation, Fudan Auto ID lab, etc. Moreover, more than 200 companies and research institutes are working on RFID middleware, data management, public information service platform and applications. Key solutions available in the market are item identification, information collection, control and processing, and near-field communication.

7.1.4 Bluetooth

It is rare to see Bluetooth enabled M2M applications in China. However, there are research institutes working on BLE (Bluetooth low energy) for sensor applications. Nokia Research Beijing is a leading research institute in Bluetooth technology. Their current focus is on BT 4.0 which enables connectivity

Key Players	Communication Module	Embedded OS	M2M enabled features	Clients of module	Applications	Maturity of market
Huawei	LTE/WCDMA/GSM/TD/EVDO/WiMax	Linux	Low energy cost, low data rate, dedicated data management (GPRS initiate, sleep and wakeup).	Operators, DTU vendor, device vendor, system integrator	Smart metering, telematics, laptop, notebook	Products available in EU market for SM and telematics; no available M2M products in China.
ZTE	WCDMA/GSM/TD/EVDO	Linux	Low energy cost, low data rate, dedicated data management (GPRS initiate, sleep and wakeup).	Operators, DTU vendor, device vendor, system integrator	Smart metering, telematics, ebook, laptop, Chongqiong tax system, ZTE logistics	Early stage in China, high cost.
SIMcom	GSM/GPRS/EDGE/WCDMA/HSDPA	Linux	Small size, high speed GPRS data transmission built for high-volume M2M applications.	Device vendor, system integrator	E-health, telematics, security monitoring,smart grid, mobile POS	Customized solution and ODM service with limited device portfolio.
Fibocom	GSM/GPRS/UMTS/GPS	Linux	Not known	Device vendor	Telematics (preparing), electric automobile charging infrastructure, smart grid, wireless POS, E-health	Mainly provide 2G and GPS modules, supplier of M2M wireless communication and LBS(location-based service) solutions

Figure 7.3: Comparison of SIM modules manufacturers.

between small devices and the mobile phone providing a complete and seamless data chain from sensors to Web services.

7.2 Data Connectivity

Data connectivity plays a key role in bridging sensing networks with carrier networks. After scouting China's market, we categorize our interested solutions into two domains: 2G/3G SIM modules and M2M gateway.

7.2.1 2G/3G SIM Modules

With the rapid growth of 3G wireless networks in China, it is straightforward and relatively low cost to develop SIM card based M2M applications and therefore the dedicated communication modules will play an important role in China's M2M development. Figure 7.3 gives an overview of key Chinese players who are capable of providing SIM modules for M2M applications.

Huawei

Huawei primarily works with 3rd party device vendors or system integrators on delivering M2M solutions to operators in specific projects. They are focusing on two target M2M markets: 1) Smart metering — SIM module and hub/DTU (downlink ZigBee, uplink GPRS) for UK smart energy. 2) Telematics — Service platform with China Unicom. Huaweis modules make up 50% of the post-install market (about 2 million units shipped for telematics in 2010.

ZTE

Module products are mainly provided for three kinds of customers: data transport utility (DTU) vendor, device vendor, and system integrator. Available products include telematics with Ford Motor, ebook with Elisa. It is also used in Chongqing tax system and ZTEs own logistics.

SIMcom

SIMcom is dedicated to wireless module solutions using multiple communication technologies such as GSM/GPRS/EDGE/WCDMA/HSDPA and provide standard modules to device vendors and system integrators for M2M, wireless local loop (WLL), notebook, GPS, etc. Partners include Qinghuatongfang for China Mobile. They also provide customized solutions and ODM services with a limited device portfolio, such as industrial modem/gateway, DTU, tracker, etc., for China Mobile, Vodafone tracker RFQ, and T-Mobile.

Fibocom

Fibocom produces wireless GSM/GPRS/UMTS modules and GPS modules. They have complete testing methods for M2M modules. They started to collaborate with Motorola in 2000 (distributor/service center/ODM) and also have a partnership with EMS (a Chinese express company) and BYD (a Chinese auto company). Overall, we conclude that M2M SIM modules from Chinese manufacturers are primarily shipped to the overseas M2M market. The M2M market in China is not mature and only has a few applications on a very small scale.

Chapter 8

Machine-to-Machine Communications for CPS

Zhengguo Sheng

University of Sussex, U.K.

Hao Wang and Daqing Gu

France Telecom Orange Labs, China

CONTENTS

Besides networking technology, the Machine-to-Machine Communications is also one of the enabling technologies of CPS, and in this chapter we focus on this topic. Specifically, we are looking at the open technologies such as IPv6 based solution that can be integrated into CPS and enable wireless sensor communications.

8.1 Introduction

Technically speaking, current CPS solutions can be categorized as non-IP-based and IP-based solutions. Most off-the-shelf solutions belong to the former, especially for some well-known standard alliances, such as ZigBee [174], Z-Wave [175], INSTEON [176], and WAVE2M [177], etc. However, most of these non-IP solutions are isolated within their own verticals, which hinders the CPS development due to the incompatible nature across heterogeneous communication systems.

Motivated by the fact that the TCP/IP protocol is the de facto standard for computer communications in today's networked world, IP-based solutions could be the future for CPS networks. In order to tackle the technical challenges, such as extensive protocol overheads against memory and computational limitations of sensor devices, Internet Engineering Task Force (IETF) takes the lead in standardizing communication protocols for resource constrained devices and developing a number of Internet protocols, including Routing Protocol for Low Power and Lossy Network (RPL) [178] and Constrained Application Protocol (CoAP) [179], etc. Besides, IP Smart Object Alliance (IPSO)[180] also actively promotes IPv6 embedded devices for M2M applications. Although it is still in its early stage to be commercialized, there is already a significant number of IP-based WSN solutions as demonstrated by a growing number of products and systems.

We are looking at open technologies that can be integrated into CPS and the IPv6-based solution is a promising one. In order to well maintain sensor devices as well as facilitate CPS application development, for example, monitoring the performance of sensor devices or sending commands to a sensor node, CPS authorities should be able to provide a reliable and efficient way to remotely monitor and control WSN without consuming significant resources. We take an approach based on the Representation State Transfer (REST) paradigm [77] in which a lightweight Web server can be embedded in resource constrained sensor devices. In essence, the proposed method is not only to integrate CPS devices into the network, but also to connect them to the "Web."

The following summarizes our contributions and key results:

- We implement the full IPv6 protocol stack on wireless sensor nodes to enable wireless connectivity among sensor devices. Specifically, the 6LowPAN/IPv6/RPL/UDP/CoAP protocol stack has been deployed on IEEE 802.15.4 radio platform.
- We integrate IEEE 802.15.4 connectivity into an open-platform gateway and implement the HTTP-CoAP proxy implementation to the OpenWrt, an open and Linux-based operating system of the gateway, to realize remote access from an ordinary IP terminal to IPv6 sensor devices.
- We propose two alternative access methods to enable REST-based applications with sensor devices. In the direct access method, the user can directly visit any sensor device by sending a CoAP request. Whereas for the proxy access method, the user can use the normal HTTP method to visit sensor devices, the gateway needs to help convert the HTTP request to the CoAP request and vise versa.

The remainder of this chapter is organized as follows. A survey of related works is provided in Section 8.2. The RESTful protocol stack used in WSN is introduced and analyzed in Section 8.3. The prototype implementation of the remote access schemes is presented in Section 8.4 and performance evaluation results are shown in Section 8.5. Finally, concluding remarks are given in Section 8.6.

8.2 Related Works

Recent technology trends in the Web Services (WS) are primarily separated as Big Web Services (or WS-*) and RESTful Web Services. Cesare et al. in [181] compare these two architecture choices and argue that the RESTful WS can create a loosely coupled system which is better suited for simple and flexible integration scenarios, whereas WS-* can provide more advanced quality-of-service for enterprise level usages.

More recent works are dedicated to developing REST-style CPS systems to enable easy access from application servers to wireless sensor devices, since the REST-style device would not require any additional API or descriptions of resources/functions. REST, a lightweight Web service implementation, is a general design style of Internet resource access protocol. It provides a design concept in which all the objects in the Internet are abstracted as resources. Each resource corresponds to a unique identity. Through a general interface, all the operations on resources do not change the identity as they are stateless.

REST style can make applications as sharable, reusable, and loose coupling services. The uniform operation and interaction mechanisms on resources can help developers or decision makers to quickly react to market changes.

Weijun et al. in [182] propose an adaptation layer to integrate the RESTful Web service infrastructure which can enable connectivity of embedded devices with mobile Internet applications. Vlad in [183] proposes a resource discovery mechanism based on RESTful principles that enables a plug and play experience in the Web of Things. Dominique et al. in [184] and [185] also propose a RESTful mechanism to integrate wireless energy monitors with application servers to build mashup applications. However, most of the embedded devices in the above literature are not IP based, which means that a multiprotocol translation gateway is needed. As discussed in [11], the network protocol translation can bring more complexity than just a packet format conversion, which usually involves semantics translation between different mechanisms and logic for routing, quality of service, and security, etc.

There are recent papers focusing on the implementation of IPv6 protocol stacks on various hardware platforms. Thomas et al. in [186] demonstrate an intelligent container testbed where the CoAP protocol is implemented on the embedded operating system TinyOS [187]. Moreover, a couple of other implementations of CoAP are also available on the Contiki platform [188, 189, 190]. However, most of these cases are only for the purpose of connectivity evaluations on different operation platforms and usually assume that a virtual gateway, which is usually an IEEE 802.15.4 USB dongle connected to a PC, is mounted as a root node to collect upstream packets from leaf nodes.

Different from the above works, our contribution in this chapter is that we consider IPv6 protocol implementation on sensor devices as well as an integrated gateway solution to allow any normal Internet devices (e.g., PC and smartphone) to visit an IPv6 sensor device. Specifically, we integrate real-world things into the existing Web by turning real objects into RESTful resources that can be retrieved directly over HTTP.

8.3 A RESTful Protocol Stack for WSN

We employ the IPv6-based protocol stack for wireless sensor networks and some IP protocols developed for resource constrained networks are introduced as follows.

8.3.1 6LoWPAN

Since the beginning of IETF research on CPS related technologies, IPv6 has been selected as the only choice to enable wireless communication. Its key features such as universality, extensibility, and stability, etc., have attracted a lot of attention and will be the de facto solution for future Internet technology.

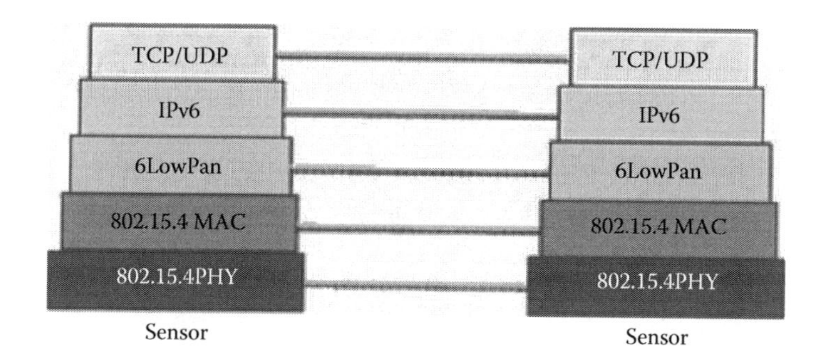

Figure 8.1: The position of 6LowPAN in the IPv6 protocol stack.

In order to enable IP connectivity in resource constrained sensor networks, the IPv6 over low-power WPAN (6LowPAN) working group is established and works on protocol optimization of IPv6 over networks made of IEEE 802.15.4 [191]. Specifically, the 6LoWPAN protocol discusses how to apply IPv6 to the MAC layer and PHY layer of IEEE 802.15.4.

In fact, there are two key challenges to running IPv6 over the IEEE 802.15.4 network. On the one hand, considering that the maximum frame size supported by IEEE 802.15.4 is only 127 bytes and there are significant header overheads occupied by layered protocols (e.g., MAC layer header, IPv6 header, security header and transmission layer), the payload size available for the application layer is very limited. On the other hand, since the minimum value of maximum transmission unit (MTU) specified by IPv6 is 1280 bytes (RFC 2460), if MTU supported by the under layer (i.e., IEEE 802.15.4) is smaller than this value, the data link layer must fragment and reassemble data packets. In order to address these issues, 6LoWPAN designs an adaptation layer right above the data link layer to tail the IPv6 packet into small pieces required by the under layer. Moreover, 6LoWPAN specifies stateless compression methods of IP header in order to reduce the overhead of IPv6. The position of 6LoWPAN in the IPv6 protocol stack is shown in Figure 8.1.

It is worth noting that the fundamental of header compression methods is to remove the redundant information from the header by using compression encoding schemes. Although the IPv6 header takes 40 bytes, most information bits can be compressed in the network sensing layer. The compression methods for each field of IPv6 header are as follows:

1. Version (4 bits): The value is 6. It can be omitted in the IPv6 network.

2. Traffic Class (8 bits): It can be compressed by compression encoding methods.

3. Flow Label (20 bits): It can be compressed by compression encoding methods.

4. Payload Length (16 bits): It can be omitted because the length of IP header can be obtained through the payload length field in the MAC header.

5. Next Header (8 bits): It can be compressed by compression encoding methods if the next header is assumed to be one of UDP, ICMP, TCP, or extended header.

6. Hop Limit (8 bits): The only field that cannot be compressed.

7. Source Address (128 bits): It can be compressed by omitting prefix or IID (interface identifier).

8. Destination Address (128 bits): It can be compressed by omitting prefix or IID.

In order to implement the stateless compression on IPv6 header, 6LoWPAN group specifies two compression algorithms: LOWPAN_HC1 (RFC4944) [192] and LOWPAN_IPHC (RFC6282) [193]. HC1 algorithm is applicable to networks using a link-local address. The prefix of node's IPv6 address is fixed as FE80::/10 and IDD can be obtained by MAC address. Since this algorithm cannot efficiently compress global/routable address or broadcast address, it cannot be used to connect the 6LoWPAN with the Internet. LOWPAN_IPHC, however, is proposed to improve the efficiency of compressing routable address.

Both LOWPAN_HC1 and LOWPAN_IPHC define an 8-bit dispatch field after the MAC header. Its values shown in Table 8.1 determine the specific format of type-specific header and algorithm. For example, if the first 8 bits is 01000010, the following field is the header corresponding to LOWPAN_HC1 algorithm; if the first 3 bits is 011, the following field is the header corresponding to LOWPAN_IPHC algorithm.

Dispatch field is immediately followed by the type-specific header which consists of some indicating bits. The indicating bits indicate a specific compression scheme for IPv6. Readers can refer to RFC4944 for more details.

In addition to stateless IPv6 header compression, 6LoWPAN also develops other relevant standards including the scheme supporting mesh routing, simplified IPv6 neighbor discovery protocol, use cases and routing requirements. In summary, the 6LoWPAN working group is the fundamental of IETF on CPS communications; its contributions significantly promote the establishment and research work of other working groups.

8.3.2 RPL

IETF Routing over Lossy and Low-power Networks (RoLL) working group was established in February 2008. It focuses on the routing protocol design and is committed to standardizing the IPv6 routing protocol for lossy and low power networks (LLN). Its tasks start with the routing requirements of

Table 8.1: 6LoWPAN Dispatch Field

Type	Header type
00 xxxxxx	NALP - Not a LoWPAN frame
01 000001	IPv6 - Uncompressed IPv6 Addresses
01 000010	LOWPAN_HC1 - LOWPAN_HC1 compressed IPv6
...	Reserved
01 010000	LOWPAN_BC0 - LOWPAN_BC0 broadcast
...	Reserved
01 XXXXXX	IPv6 header compressed by LOWPAN_IPHC
01 000000	ESC: indicating more headers to come
10 xxxxxx	MESH - Mesh Header
11 000xxx	FRAG1- Fragmentation Header (first)
11 100xxx	FRAGN - Fragmentation Header (subsequent)

various application scenarios. So far, the routing requirements of four application scenarios have been standardized, i.e., home automation (RFC5826), industrial control (RFC5673), urban environment (RFC5548), and building automation (RFC 5867).

In order to develop suitable standards for lossy and low-power networks, RoLL first provides an overview of existing routing protocols for wireless sensor networks. The literature [194] analyzes the characteristics and shortcomings of the relevant standards and then discusses the quantitative metrics for constructing routing in the routing protocol. RFC6551 [195] introduces two kinds of quantitative metric: node metrics including node state, node energy, and hop count, and link metrics including throughput, latency, link reliability, expected transmission count (ETC), and link colour object. In order to assist dynamic routing, nodes can design objective function to determine the rule to select a path based on the quantitative metrics.

Based on the results of routing requirements and quantitative static link metrics, RoLL develops routing protocol for LLN (RPL), i.e., RFC6550 [196]. RPL supports three kinds of traffic flow including point-to-point (between devices inside the LLN), point-to-multipoint (from a central control point to a subset of devices inside the LLN), and multipoint-to-point (from devices inside the LLN toward a central control point). RPL is a distance-vector routing protocol, in which nodes construct a Directed Acyclic Graph (DAG) by exchanging distance vectors. Through broadcasting routing constraints, DAG root node (i.e., central control point) filters out the nodes that do not meet the constraints and selects the optimum path according to the metrics.

8.3.3 CoAP

Constrained Application Protocol (CoAP), specified by IETF CoRE working group [179], is a specialized Web transfer protocol for resource constrained

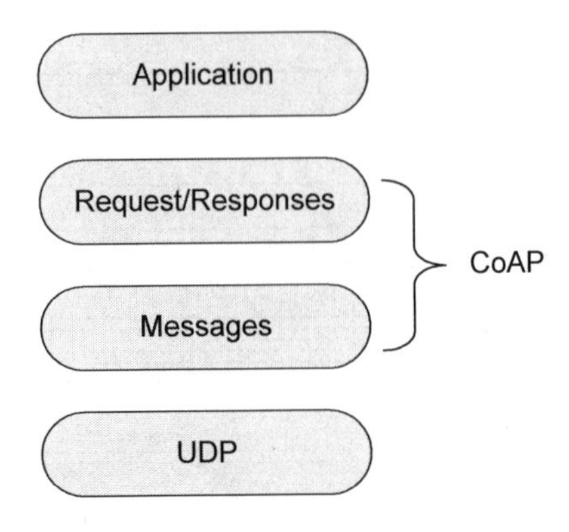

Figure 8.2: CoAP protocol stack.

nodes and networks. CoAP conforms to REST style. It abstracts all the objects in the network as resources. Each resource corresponds to a unique Universal Resource Identifier (URI) from which the resources can be operated stateless, including GET, PUT, POST, DELETE, and so on.

Strictly speaking, CoAP is not a HTTP compression protocol. On the one hand, CoAP realizes a subset of HTTP functions and is optimized for constrained environment. On the other hand, it offers features such as built-in resource discovery, multicast support, and asynchronous message exchanges.

Unlike HTTP, CoAP adopts datagram-oriented transport protocols, such as UDP. In order to ensure reliable transmission over UDP, CoAP introduces a two-layer structure which is shown in Figure 8.2. The messaging layer is used to deal with asynchronous interactions using UDP. Specifically, there are 4 kinds of CoAP message:

1. Confirmable (CON): ACK is needed.
2. Non-confirmable (NON): ACK is not needed.
3. Acknowledgment (ACK): To represent that a confirmable message is received.
4. Reset (RST): To represent that a confirmable message is received but can't be processed.

The Request/Response interaction layer is used to transmit resource operation requests and the request/response data. As a summary, CoAP has the following features:

- Constrained Web protocol fulfilling M2M requirements.
- Asynchronous message exchanges.
- Low header overhead and parsing complexity.

- URI and content-type support.
- Simple proxy and caching capabilities.
- Built-in resource discovery.
- UDP binding with optional reliability supporting unicast and multicast requests.
- A stateless HTTP-CoAP mapping, allowing proxy to provide access to CoAP resources via HTTP in a uniform way and vice versa.

8.3.4 HTTP-CoAP Protocol Implementation

Applying REST-style network structure in WSN can largely facilitate connection between WSN and the Internet. By applying CoAP protocol on wireless sensor devices, Internet services can access wireless sensor networks as resources directly or via gateway as a proxy. Basically, there are two methods to enable remote access from an Internet client to a sensor device.

8.3.4.1 Direct Access

Direct access means that the Internet users access WSN through a gateway which only implements protocol conversion between the IPv6 network layer and 6LoWPAN, but does not process the upper layers protocols (e.g., CoAP). As an example shown in Figure 8.3 (a), a sensor node in WSN can be visited through an IPv6 address and the gateway only needs to implement conversion between IPv6 and 6LoWPAN, which significantly reduces the processing overhead.

8.3.4.2 Proxy Access

Proxy access means that Internet users access WSN through a proxy which can convert an incompatible data format from outside networks into a WSN compatible data format. For example, in our case, the proxy can have functions of protocol conversion from a HTTP request to a CoAP request, and vice versa, payload conversion, and blockwise segmentation of large data packet (e.g., image), etc.

The advantage of this method is that current Internet services can easily access WSN resources without any changes, because of the existing proxy gateway. Moreover, since the low power sensor mode cannot serve TCP requests efficiently, the proxy mechanism can buffer and process the requests to avoid TCP time out. However, the protocol conversion increases the complexity of the gateway and thereafter affects communication efficiency. Figure 8.3 (b) illustrates the protocol conversion between HTTP and CoAP via gateway.

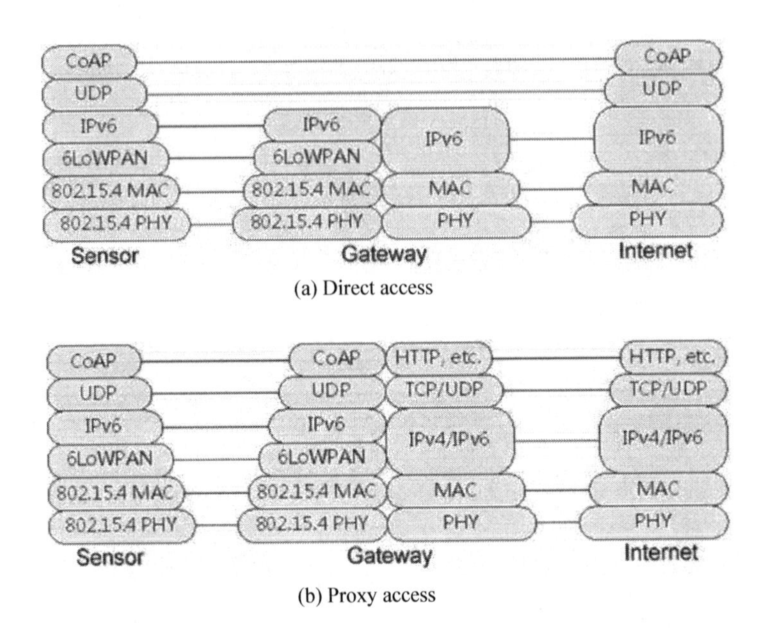

Figure 8.3: Direct access vs. proxy access.

8.4 Prototyping Implementation

In this section, we present our prototyping system to illustrate the implementation of the RESTful access methods to IPv6 wireless sensor devices. Specifically, we choose IPv6 enabled wireless sensor devices as the representative of the embedded devices. A RESTful gateway supports both Wi-Fi interface and IEEE 802.15.4 communication. The Web resources located on sensor devices are accessible through the RESTful APIs. The system architecture is shown in Figure 8.4, where a personal computer acts as a client to retrieve sensor resources via the RESTful gateway.

8.4.1 Sensor Node

We deploy wireless sensor devices to monitor air temperature and humidity, detect movement, and take photos. All these sensor platforms are equipped with the same ATmega1284P MCU and AT86RF231 transceiver to support 250kbps, 2.4GHz, and IEEE 802.15.4 radio. To support IPv6 connectivity, all the sensor devices are running Contiki v2.6 operating system and implement 6LowPAN, IPv6, and RPL protocols based on IEEE 802.15.4. The Web service running on the sensor devices relies on the application protocol CoAP and uses version 06 of the draft. A snapshot of a sensor platform is illustrated in Figure 8.5 and the detailed technical specifications are shown in Table 8.2.

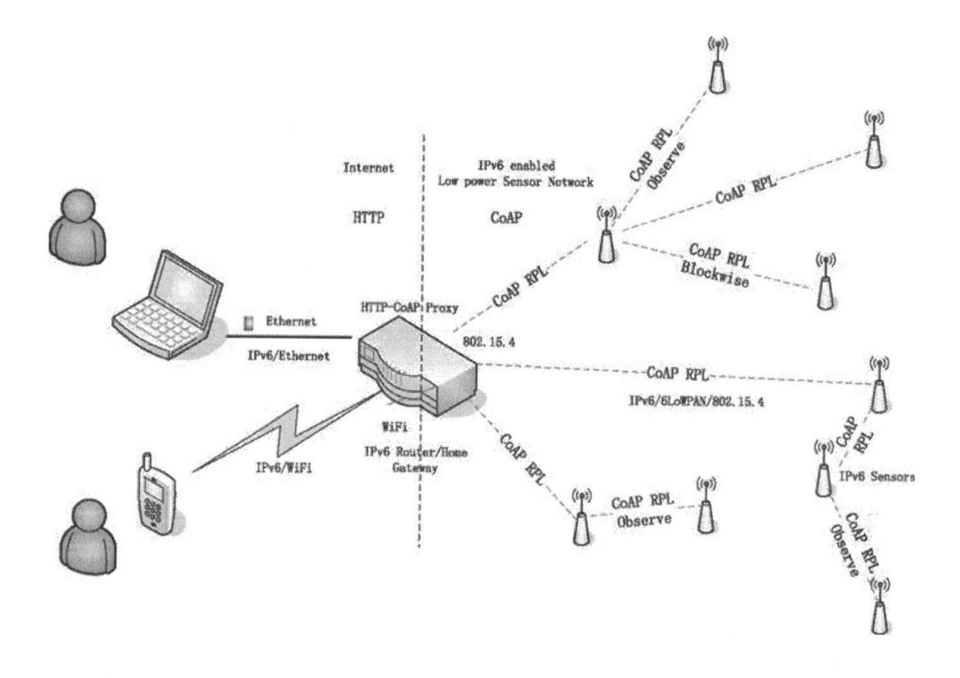

Figure 8.4: System architecture.

Figure 8.5: A snapshot of a sensor platform.

8.4.2 RESTful Gateway

To ease the access from Internet applications to sensor resources, especially for those Internet users who cannot speak CoAP, we integrate IEEE 802.15.4 connectivity into an open-platform gateway and port the HTTP-CoAP proxy implementation to the OpenWrt, the operation system of the gateway, to

Table 8.2: Technical Specifications of Sensor Device

	Parameters	Note
CPU Performance		
Internal storage	128KB	
External storage	16KB	
EEPROM	4KB	
Serial communication	UART / USART	TTL transmission level
A/D converter	10-bit ADC	8 channels, 0-3V input
Other interfaces	Digital I/O, I2C,SPI	
Maximum current	18mA	Work mode
	2uA	Sleep mode
RF Transceiver		
Frequency band	2400-2485MHz	ISM global free band
Data rate	250Kbps/ 1000Kbps/ 2000Kbps	
RF power	3.2 dBm	
Receiving sensitivity	−104 dBm	
Adjacent channel suppression	36 dBc	+5M Channel bandwidth
	34 dBc	-5M Channel bandwidth
Outdoor transmission	$\geq$ 300m	
Indoor transmission	$\geq$ 10m	
Maximum current	12mA	Receiving mode
	14mA	Tx -3dBm
Extended interface	51 pins	

realize remote access from an ordinary IP terminal to an IPv6 sensor device. Figure 8.6 gives the hardware architecture of the RESTful gateway. The technical specifications are also provided in Table 8.3.

HTTP-CoAP (HC) proxy provides translation and mapping between HTTP and CoAP protocol. CoAP can be directly mapped to HTTP, because CoAP actually implements a subset of HTTP functions. The mapping is performed only at the request/response layer of the CoAP protocol and is invisible to the messages layer. There are two kinds of mapping: CoAP-to-HTTP and HTTP-to-CoAP. In our case, we only realize the HTTP-to-CoAP mapping which is implemented by specifying CoAP-URI as the request address or transmitting HTTP request to the HTTP-CoAP proxy. It is worth noting that, compared to CoAP-to-HTTP mapping, the HTTP-to-CoAP mapping is more complex, since it is necessary to determine whether to ignore the content or report an error by checking unsupported HTTP request methods, response codes, content-types, and options.

In our prototype gateway, the HC proxy is implemented based on libcoap [197] which is an open-source C-Implementation of CoAP and conforms to

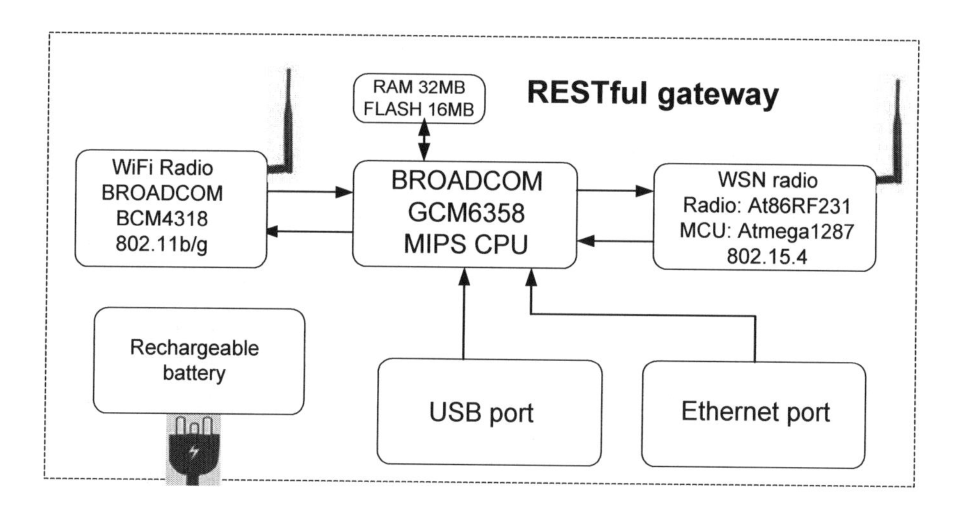

Figure 8.6: Hardware architecture of gateway.

Table 8.3: Technical Specifications of Gateway

	Parameters	Note
CPU frequency	300MHz	
RAM	32MB	
Flash	16MB	
Serial communication	UART / USART	TTL Transmission level
A/D converter	10-bit ADC	8 channels, 0-3V input
USB HOST	2	
RJ45	4	
WiFi	1	IEEE 802.11abg
OS	OpenWrt	v12.09-beta2
Protocol	IPv6, IPv4	

GPL v2 or higher licenses. The version we adopted is 2.1.0 with support of coap-06.

The interaction process of the HC proxy is shown in Figure 8.7. Specifically, for each of the HC proxy layers, we have the following implementations:

8.4.2.1 libcoap Layer

libcoap implements the CoAP messages layer based on UDP. It defines CoAP message structure and methods to operate CoAP messages.

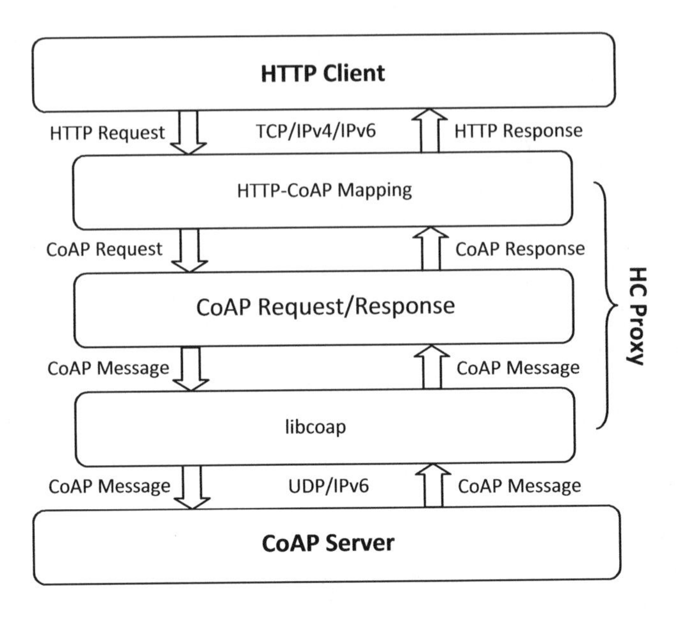

Figure 8.7: Interaction process of HC proxy.

8.4.2.2 CoAP Request/Response Layer

CoAP Request/Response layer encapsulates the data structure and methods relevant to CoAP requests and responses. It is to transmit CoAP requests in the form of CoAP messages through the messages layer and generate CoAP response based on received CoAP messages. Because the CoAP messages layer adopts unreliable UDP, certain issues need to be solved in order to implement a reliable transmission, including CoAP message acknowledge, message retransmission for timeout, message process, asynchronous message process, and segmented message process, etc. The following code header is provided to illustrate the implementation of CoAP Request/Response.

```
1   /* request.h */
2
3   typedef struct {
4     unsigned char msgtype;
5     method_t method;
6     coap_list_t *optlist;
7     str proxy;
8     unsigned short proxy_port;
9     str payload;
10    int ready;
11    char lport_str[NI_MAXSERV];
12    coap_uri_t uri;
13    int flags;
```

```
14    coap_block_t block;
15    unsigned int wait_seconds;    /* default timeout in
         seconds */
16    coap_tick_t max_wait;         /* global timeout (
         changed by set_timeout()) */
17    unsigned int obs_seconds;     /* default observe time
         */
18    coap_tick_t obs_wait; /* timeout for current
         subscription */
19  } coap_request_t;
20
21  void coap_request_method(coap_request_t *request, char
       *arg);
22  void coap_request_uri(coap_request_t *request, char *
       arg);
23  int coap_request_proxy(coap_request_t *request, char *
       arg);
24  void coap_option_content_type(coap_request_t *request,
       char *arg, unsigned short key);
25  int coap_option_blocksize(coap_request_t *request, char
       *arg);
26  void coap_option_subscribe(coap_request_t *request,
       char *arg);
27  void coap_option_token(coap_request_t *request, char *
       arg);
28  int coap_send_request(coap_request_t *request, void *
       context);
29
30  void coap_init_request(coap_request_t *request);
31  void coap_register_request_handler(int (*handler)(
       coap_pdu_t *pdu, void *context));
32  void coap_register_request_data_handler(int (*handler)
       (const unsigned char *data, size_t len, void *
       context));
```

8.4.2.3 HTTP-CoAP Mapping Layer

It is to implement mapping from HTTP requests to CoAP requests and vice versa. When converting a HTTP request to a CoAP request, the HC proxy needs to convert the HTTP request method, URI, header/option, and payload, respectively. If a proxy encounters an error, it has to generate the corresponding error response. The C function defined for handling the HTTP-CoAP mapping is also provided as follows.

```
1  int coap_response_map_code(int code);
2  char* coap_response_map_content_type(int content_type);
3
```

```
4 BOOL coap_proxy_handler(SOCKET localwebuser, char *
      szLineBuffer, int nLineBuffer)
```

8.5 Performance Evaluation

In this section, we provide evaluation results of the prototype system. Especially, we evaluate the performance from two layers: the routing layer where the round trip times (RTTs) and packet loss rate of multi-hop transmission in wireless sensor networks are measured and the application layer where Web resources of sensor devices are retrieved using RESTful methods.

8.5.1 *System Configuration*

Our prototype system is composed of three different sensor devices, one HC proxy gateway, and one PC for initiating tests. In order to ease the setup of WSN in a multi-hop fashion, we manually assign IPv6 addresses for the sensor devices as follows:

Camera sensor	2001:2::19
Humidity & temperature sensor	2001:2::14
Approach detecting sensor	2001:2::16

We deploy the prototype system in an open office area. The HC proxy gateway and sensor devices are connected wirelessly via IEEE 802.15.4 and using channel 26. The PC client is connected to the gateway through the Wi-Fi channel. The network topology is built with a maximum number of 2 hops, where the camera sensor and humidity/temperature sensor are directly connected to the gateway with one hop distance, and the approach detecting sensor is the leaf node of the humidity/temperature sensor and it is two hops away to the gateway. The distance of each hop is around 10 meters. Figure 8.8 provides the network topology of the prototype system.

8.5.2 *RTTs and Packet Loss Evaluations of RPL Routing*

To cope with large-scale deployment and low cost requirements of WSN, wireless sensor networks should be capable of forming multi-hop transmissions among peer sensor devices. In this evaluation, the RTTs and packet loss rate in single-hop and multi-hop scenarios are measured by using RPL routing protocol. After the setup of the system, we use the simple ping commands to evaluate the RTTs from the PC client to the humidity/temperature sensor and approach detecting sensor, respectively. The payload size for each transmission packet is 32 bytes and the RTTs results are averaged over 100

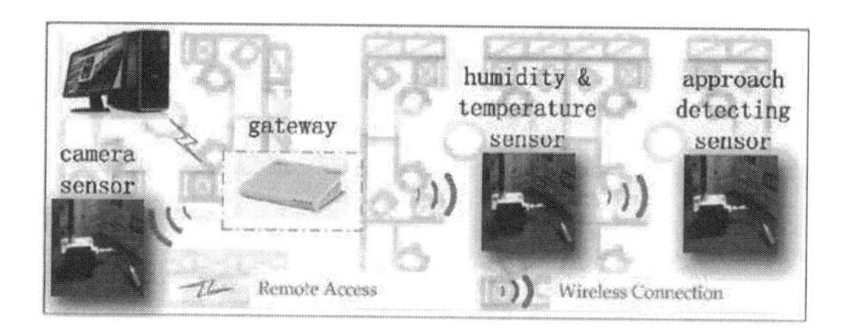

Figure 8.8: Network topology of the prototype system.

measurements. Figure 8.9 (a) shows the routing table via SSH client. As can be seen from Figure 8.9 (b), for one-hop transmission, the average RTTs and packet loss rate are 24 ms and 18%, respectively. When the routing extends to two hops, the results as shown in Figure 8.9 (c) are degraded to 43 ms average RTTs and 20% packet loss.

There are several factors that may cause the performance losses. Since the prototype system is exposed in a highly polluted environment with more than 10 Wi-Fi access points jamming the IEEE 802.15.4 radio, some of the packets may be interfered with and overwhelmed by nearby Wi-Fi signals. Furthermore, the hardware constraints and protocol configurations also affects the routing performance.

8.5.3 RESTful Method to Retrieve Sensor Resources

To drive CPS applications, we initiate a trial by 'GET' a JEPG picture from the camera sensor device. Specifically, we use both proxy access and direct access methods to retrieve the sensor data via the gateway. Figure 8.10 (a) shows the proxy access result by sending a HTTP GET request along with the URI `http://[2001:2::19]/camera`. The HC proxy then converts the HTTP request to CoAP request and forwards the request to the camera sensor. In contrast, Figure 8.10 (b) shows the direct access result by sending a CoAP request `coap://[2001:2::19]:5683/camera` directly from the CoAP browser [198] on the PC. It is worth noting that since the picture takes about 27 kBytes, which exceeds the payload size defined by the CoAP client, the CoAP protocol adopts the blockwise transfer by dividing the response into 64-byte blocks in such a way that the Web server can handle each block transfer separately, with no need for a connection setup or other server-side memory of previous block transfers. In summary, both methods show acceptable performances.

```
Routes [6 max]
2001:2::14/128 (via fe80::14)
2001:2::16/128 (via fe80::14)

---------

Connected to 192.168.1.1
```

(a) Routing table

```
Ping statistics for 2001:2::14:
    Packets: Sent = 100, Received = 82, Lost = 18 (18% loss),
Approximate round trip times in milli-seconds:
    Minimum = 16ms, Maximum = 126ms, Average = 24ms
```

(b) RRTs and packet loss rate from one-hop sensor device

```
Ping statistics for 2001:2::16:
    Packets: Sent = 100, Received = 80, Lost = 20 (20% loss),
Approximate round trip times in milli-seconds:
    Minimum = 37ms, Maximum = 212ms, Average = 43ms
```

(c) RRTs and packet loss rate from two-hop sensor device

Figure 8.9: Routing table, RTTs, and packet loss evaluations.

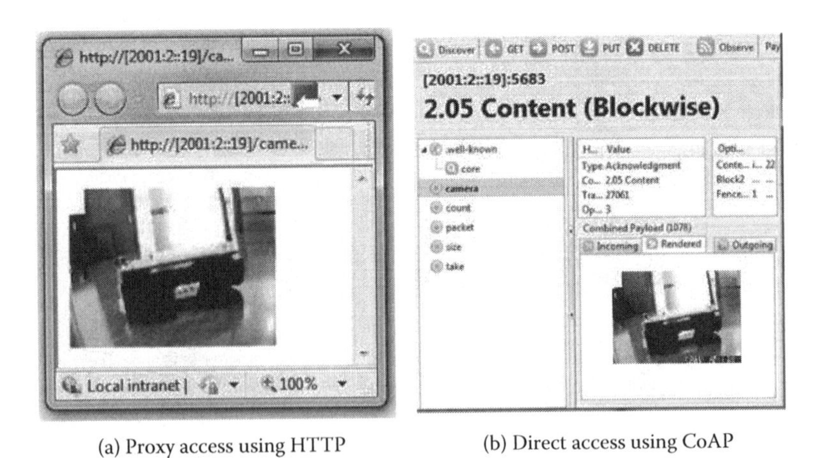

(a) Proxy access using HTTP

(b) Direct access using CoAP

Figure 8.10: HTTP vs. CoAP methods.

8.6 Summary

We have implemented the 6LowPAN/IPv6/RPL/CoAP protocol stack on IEEE 802.15.4 radio platform to enable wireless sensor communications. Furthermore, by integrating IEEE 802.15.4 connectivity and HTTP-CoAP proxy into the open-platform gateway, we have realized remote access from a PC to an IPv6 sensor device. Through the performance evaluations, we have shown the IP-based solution is promising to drive IOT development. In future work, we plan to design a more robust and reliable IP solution for CPS. Especially, how to deploy large-scale networks with decent performance is a critical issue and we need to continue to optimize both hardware and software implementations. Moreover, other issues, such as device management and control of sensor devices, can also be explored via RESTful methods.

Chapter 9

Mobile Cloud Computing for CPS

Chi Harold Liu

Beijing Institute of Technology, China

CONTENTS

Recently, the evolution of mobile cloud computing (MCC) has attracted significant attention from both academia and industry. Similar to networking and machine-to-machine communications, it is more and more considered as a critical method to power and boost the CPS. In this chapter, we will discuss state-of-the-art MCC research activities that aim to enhance the capabilities of resource-constrained smart devices in CPS sensory environments.

9.1 Introduction

The underlying motivation of mobile cloud computing (MCC) is based on the desire to overcome the resource constraints on smart devices, such as computation power, memory, and storage capacity to satisfy the ever increasing quality-of-service (QoS) requirements of diverse CPS applications on-the-move. Typical QoS parameters include, but are not limited to, throughput, user experienced latency, service reliability and availability, etc. MCC is originally rooted in interdisciplinary research of mobile (ubiquitous) computing and cloud computing. Existing research activities try to cross the disciplines' boundaries by applying cloud computing solutions and techniques to mobile applications, or incorporating mobile features when constructing new cloud services. However, the immense and complex information involved in mobile cloud applications for CPS demands new transdisciplinary research to better understand the natures and principles of MCC. Toward this end, MCC, which has recently attracted significant attention from both academia and industry, is more and more considered as a critical method to power and boost the CPS, enabled by resource-constrained smart devices.

In a CPS sensory environment, the term *mobile cloud* can be interpreted in three perspectives that ultimately form three types of MCC-enabled CPS architectures, as illustrated in Figure 9.1. They are: (a) `Internet Cloud:` that provides leverages to individual smart devices, (b) `Device Cloud:` that

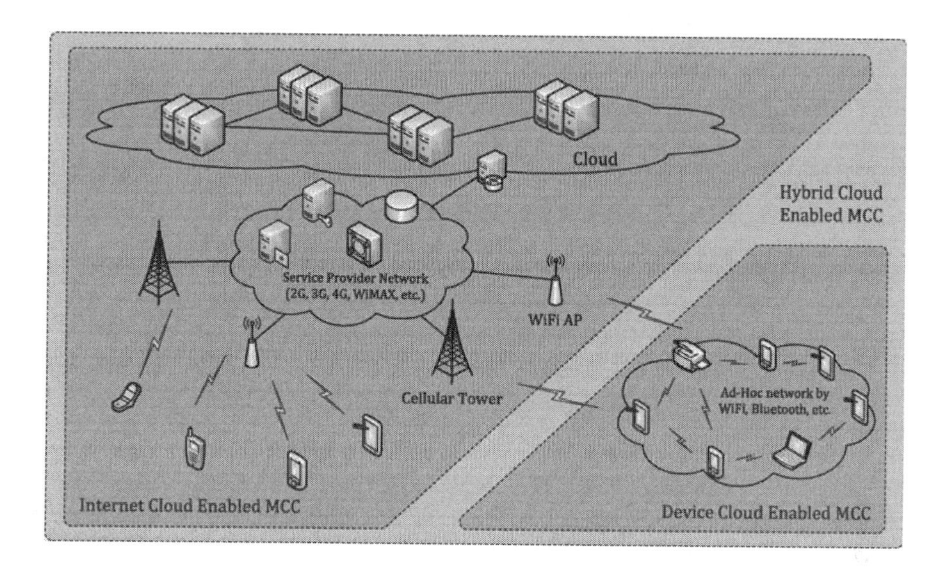

Figure 9.1: Three MCC-enabled CPS architectures: Internet Cloud, Device Cloud, and Hybrid Cloud.

consists of a group of physically co-located smart devices, and (c) `Hybrid Cloud`: in which an individual device can leverage the resources from both `Internet Cloud` and `Device Cloud` simultaneously. In the first architecture, the `Internet Cloud` is a relatively static infrastructure (usually at the back-end) that provides computation or storage resources via the Internet to the end devices. On the other hand, a set of smart devices themselves, especially those physically located in the proximity, can form a `Device Cloud` locally and share computation or storage resources among each other to achieve certain sensing tasks in a CPS sensory environment. To the best of our knowledge, most of the MCC interests in the field are targeting the challenges from the first perspective aforementioned. However, we do not limit the selection of CPS applications to the former cases only, because we reckon the challenges faced by the dynamic and ad hoc mobile cloud are fundamentally different when considering the underlying CPS architecture and unpredictable mobility patterns of those devices. Furthermore, the opportunities and benefits of evolving the interactions of two kinds of clouds, such as `Hybrid Cloud`, are most promising and should not be neglected for CPS sensory environments.

In this chapter, we survey the state-of-the-art MCC research activities that aim to enhance the capabilities of resource-constrained smart devices in CPS sensory environments by means of either employing existing Internet-based cloud computing infrastructures or locally constructing an ad hoc device cloud. Our contribution is threefold: (a) three types of MCC-enabled CPS architectures are proposed, reviewed, and extensively discussed, (b) applicabilities of these three architectures, both existing and new, on CPS are described, and (c) their associated new research challenges and opportunities are identified.

The remainder of this chapter is organized as follows: the basic definitions of MCC are revisited in Section 9.2. We then review the associated challenges, representative solutions, and open research issues from three major categories in Section 9.3. Section 9.4 presents some promising future research directions within the context of our discussed topics. Finally, the chapter is concluded in Section 9.5.

9.2 MCC Definition

Participatory sensing focuses on the seamless collection of information from a large number of connected smart devices and provides fine-grained monitoring of environmental phenomena; thus fundamentally it can be regarded as a key enabling technology for CPS, where the only difference, if compared with the traditional understanding of a CPS sensory environment, is the employed devices. With the embedded sensors like accelerometer, gyroscope, GPS, camera, and microphones, smart devices integrated with rich media and location tracking features are making the concept of participatory sensing more feasible and realizable, by forming a ubiquitous mobile cloud and providing comprehensive information perceiving. Given the current commercial hype and the fact that there are diverse ways of cooperating with cloud computing and mobile applications, it is not surprising that there is no single widely accepted clear definition of MCC. As a result its scope may not be properly defined [199]. Cloud computing is the delivery of computing as a service rather than a product, whereby shared resources, software, and information are provided to computers and other devices as a utility (like the electricity grid) over a network (typically the Internet). It provides computation, software, and storage services that do not require end-user knowledge of the physical location and system configuration. Cloud computing research mainly focuses on how to manage computing, storage, and communication resources that are shared by multiple users in a virtualized and isolated environment.

There are **three** main different service models in the `Internet Cloud` computing environment: **(a)** *Software as a Service* (SaaS), which provides access for the end users to a specific application such as Gmail, Facebook, Microsoft Office 365; **(b)** *Platform as a Service* (*PaaS*) provides application programming interface (API) and programming environments for the application developers, (Google Application Engine (GAE), Amazon Web Services and Microsoft Azure are in this layer); and **(c)** finally *Infrastructure as a Service* (*IaaS*), provides computation and storage facilities through virtualization such as Amazon EC2 and S3.

Although there is no single widely accepted definition of MCC in the field [200], we refer to MCC as the set of techniques that are using shared resources (in different manners) to empower mobile applications. The resources can be provided in the form of computation power, storage capacity, or information content from either or both of the infrastructure based cloud and other mobile devices locally in the ad hoc, dynamic `Device Cloud`. The overall goal is to provide better QoS experience for end users with these limited resources and capacities like computation, storage, and battery.

Figure 12.1 shows the general ecosystem of MCC and some of its important players [201] including, but not limited to, public cloud providers, content and service providers, local and private cloud providers, network providers, smartphones, and sensors. Public cloud providers serve *scalable* and *elastic* high computation and storage facilities with the help of globally distributed data centers. Local and private cloud providers have limited scalability and elasticity but with high performance which can provide efficient services on the portable devices considering delay, energy consumption, local information caching, data protection, and privacy, etc. Content and service providers are providing usable services and data content for the end mobile users like news, videos, email, games, etc. They could get their required storage and servers from public and local cloud providers. Network providers provide the network infrastructures (wired or wireless) where all MCC architectural components could communicate. Finally, smart devices and sensors like iPad, smartphones, traffic cameras and sensors distributed in cities, etc., are also important components in this ecosystem of MCC.

9.3 Challenges

By forming a ubiquitous and dynamic mobile cloud and enhancing the capacity of smart devices, MCC-enabled CPS applications and services will ultimately significantly enrich the information perceiving and processing processes. However, its success heavily relies on a few key technologies, namely: (a) efficient task offloading, (b) manageable heterogeneity, and (c) sufficient security and privacy protection. In this section, we summarize a detailed taxonomy of the most creditable and recently developed solutions in Table 9.1, and identify research gaps regarding these three characteristics in practice.

9.3.1 Managing the Task Offloading

Since cloud resources in any of the three CPS architectures have considerably better computation, storage performance, and constant power supply, offloading the "heavy-lifting" to the cloud appears to be the feasible solution to overcome the resource limitations on smart devices. A significant amount of research has been conducted on the offloading techniques, where a recent systematic survey [213] indicates that, in the last fifteen years, researchers' focus has moved from making offloading feasible and making offloading decisions (e.g., whether, what and how) to the infrastructure of offloading.

In [213, 214] the authors were among the first to propose simple analytical models that help to decide whether to offload. The parameters take into

Table 9.1: Comparison of Mobile Cloud Computing Solutions

Approaches	Objectives	Partitioning	Scheduling	Heterogeneity	Security/Privacy	Pros/Cons
MapReduce [202]	Performance	Manually annotation on method level	Static	-	Trustworthy cloud assumed	Users manually mark the "offloadable" block.
Hyrax [203]	Performance	Manually annotation on method level	Static	Homogeneous hardware allowed.	Trustworthy cloud assumed	Implemented based on Hadoop/MapReduce
Cloudlets [204]	Performance	-	-	Commodity server hardwares are used.	The physical security of cloudlets is easy to safeguard.	Cloudlet-type solutions seek a compromise between distributed and centralized operation.
MAUI [205]	Battery and Performance	Manually annotation on method level	Dynamically by profiling each method and the network. History based profiling	Dynamic decisions incorporate changes in the network conditions and CPU consumption of individual methods over time.	Trustworthy cloud assumed	Users manually mark the "offloadable" block
MobiCloud [206]	Enhance Security	-	Dynamic: Stochastic method	Context-aware risk assessment	Fundamental trust model is provided.	A secure service-oriented mobile cloud framework is built to assist communications in MANETs.
Scavenger [207]	Performance	Automatically annotation on application level	Dynamic: dual-profiling scheduler	Dynamic offloading	Black-listing and white-listing approach is used to accept new mobile node.	Scavenger is the only system that considers the CPU utilization effect. Easy for programmers to use.
CloneCloud [208]	User specifiable; reduction in execution time, energy usage.	Automatically static analysis on methods and migrate threads at runtime	Dynamic: History based profiling	Dynamic offloading	Trustworthy cloud assumed	Live migration of virtual machines enables moving an entire OS and all its running applications.
ThinkAir [209]	Battery	Automatically on method level	Dynamic: History based profiling	Dynamic offloading	Trustworthy cloud assumed	Scalability is enhanced by parallelism on dynamic VMs.
MAPCloud [210]	Performance and Scalability	Automatically on method level	Dynamic: Heuristics resource allocation	Middleware broker	Trustworthy cloud assumed	Heuristics is used to optimize resource allocation.
Cuckoo [211]	Performance and Battery	-	-	Dynamic offloading	-	The program needs to be rewritten.
Carmen [212]	Enhance Connectivities	-	Comprehensive and accurate resource information from a users' MPG are used to make better decisions on connectivity.	Seamless network connectivity is proposed for individual user with multiple devices in heterogeneous networks.	Security and privacy are considered.	Focus on mobile connectivity across multiple devices and deep integration with the cloud and the network infrastructure.

consideration are: the amount of computation need to be done as w, the speed of the cloud as s_s, the speed of the mobile device m as s_m, the bandwidth between the mobile device and the cloud as B, the amount of data sent to the cloud as d_i. Then offloading improves the performance when the amount of time of executing the tasks takes longer on the mobile device:

$$\frac{w}{s_m} > \frac{d_i}{B} + \frac{w}{s_s} \Rightarrow w \times \left(\frac{1}{s_m} - \frac{1}{s_s}\right) > \frac{d_i}{B}. \tag{9.1}$$

Similarly, offloading saves energy if the energy consumption of executing the tasks is larger on the mobile device:

$$p_m \times \frac{w}{s_m} > p_c \times \frac{d_i}{B} + p_i \times \frac{w}{s_s} \Rightarrow w \times \left(\frac{p_m}{s_m} - \frac{p_i}{s_s}\right) > p_c \times \frac{d_i}{B}, \tag{9.2}$$

where p denotes mobile's power during *computing* as p_m, *communicating* as p_c, and *waiting for response* as p_i. Therefore, only tasks that require heavy computation (large w) with light data exchange (small d_i) should be considered for the purpose of both performance improvement and energy saving.

Furthermore, various other parameters should be taken into account in the decision-making process, such as migration overhead, security and privacy, and the required bandwidth and exchanged information may change for different offloading tasks. Thus, a majority of the recent research focus on making *dynamic* offloading decision at runtime [205, 215, 216, 217, 218, 219] rather than *static* [220, 221].

Before the decision to offload is made, the "offloadable" elements of code have to be identified, which can be done either manually [205] by developers or automatically [208] by partitioning tools. Memory arithmetic unit and interface (MAUI) [205] uses a combination of *virtual machine migration* (VMM) and *code partitioning*. Their main objective is to save energy while considering application execution time. Applications are offloaded from phones to local servers using WiFi or 3G. They used .NET framework for implementation. MAUI's partitioning is done at runtime and it is very dynamic. Developers annotate which methods can be offloaded and at the time of execution, if there is a local cloud (local server), MAUI decides whether or not to offload these methods. By profiling application cost like device energy usage and network characteristics, the data from the profiler is then fed into the MAUI *Solver* to decide how to partition application optimally.

To avoid the manual annotation, CloneCloud [208] automatically marks the potential "offloadable" blocks in the bytecode statically, and then dynamically determines the optimal offloading at runtime. CloneCloud uses VMM to transfer part of a mobile application to a server using 3G or WiFi wireless connections. VMM simply refers to transferring the memory image, CPU state, storage contents, and network connection from a source node to a destination node. It greatly reduces the programming effort while there is no need to rewrite the application to be adapted for different environments.

CloneCloud has a *cost model* that considers the cost involved in migration of the application into the cloud and compares the cost against running it on a mobile device. To calculate the cost it uses the combination of "*static analyzer*" and "*dynamic analyzer*" to optimally partition the mobile application.

On the one hand, *static analyzer* uses well-defined policies for placing migration and re-integration points in the code. To solve traceable optimization problems, CloneCloud used the following policies:

- *Migration* and re-integration points should be considered in the entry and exit of the methods.
- *Methods* that access *specific features* of a mobile device such as Global Positioning System (GPS) must be pinned to the device.
- *Methods* that *share native states* must be collocated at the same machine.
- *Nested* and *core-system library* migration is prevented.

On the other hand, the *dynamic analyzer* uses randomly chosen input data fed into the application to collect the execution cost on mobile device and cloud. It is used to construct a cost model for the application under different execution settings. This data is then fed into the "*Optimization Solver*" to decide which method needs to be migrated to minimize execution time and energy usage.

Rather than considering energy efficiency on individual devices, researchers [222] develop computational offloading schemes that maximize the lifetime of an entire collection of highly collaborative mobile devices. Essentially, the schemes aim to keep every mobile device in the collection alive by finding out the best approach to schedule and shuttle the computation workload among the set of mobile devices.

9.3.2 Encountering Heterogeneity

As illustrated in Figure 9.1, high heterogeneity is an inevitable feature of a CPS ecosystem since MCC aims to bring together players from multiple disciplines, such as network providers, public and local cloud providers, smart devices, and sensors. Table 9.2 demonstrates only a small fraction of the heterogeneity in the mobile devices, network connections, and the clouds. This makes an important issue on how to seamlessly orchestrate this diversity in CPS to achieve high QoS.

To achieve maximum benefits from heterogeneity and overcome its challenges, researchers have leveraged several technologies and approaches like visualization, middleware, and service oriented architecture (SOA) which are referred to as heterogeneity handling approaches in [223]. We reckon the development of a CPS middleware is one of the most promising directions to counter the challenge of heterogeneity that highly leverages the `Internet Cloud` architecture.

Table 9.2: Heterogeneity of Hardware

Mobiles	Battery	Storage	Performance
Handheld devices	Low	Low	Low
Laptops	Med	High	Med/Low
Vehicles	High	Med/Low	Med/Low

Networks	Latency	Bandwidth	Power Efficiency
2.5/3G	High	Low	Low
WiFi	Med/Low	Med/High	High
4G LTE	Low	High	Med/High

Clouds	Scalability	Performance	Security Overhead
Public	High	High	High
Private	Med/Low	High	Low
Hybrid	Med/Low	High	Med/High

MAPCloud [210] aims to improve performance and scalability of MCC by proposing a hybrid 2-tier architecture that consists of both local private cloud and public cloud. As shown in Figure 9.2, the middleware architecture maintains a registry of resources and services and resource allocation in both tiers of the cloud, and uses heuristics to optimize the resource allocation. *Mobile User Log Database*: This database (DB) contains unprocessed user data logs such as mobile service usage, location of the user, user delay in getting the service, energy consumed on user mobile device, etc. *Mobile Profile Analyzer*: This module processes mobile user Log DB. Based on the processed data it will update QoS-aware cloud service DB, space-time mobile user pattern DB, and MAPCloud Analytics DB. *QoS-Aware Cloud DB*. It contains the service lists on local and public cloud and their QoSes in different locations. It will be updated based on the mobile user statistics experience in different locations. *MAPCloud Analytics DB*: This DB contains general system performance, such as improved energy saving, price, and delay. *Mobile User Space-Time DB*: This DB contains space-time workflow pattern of the mobile users. It contains the user service usage in space and time. *Admission Control and Scheduling*: This module is responsible for optimally allocating services to mobile users based on a heuristic called CRAM (Cloud Resource Allocation

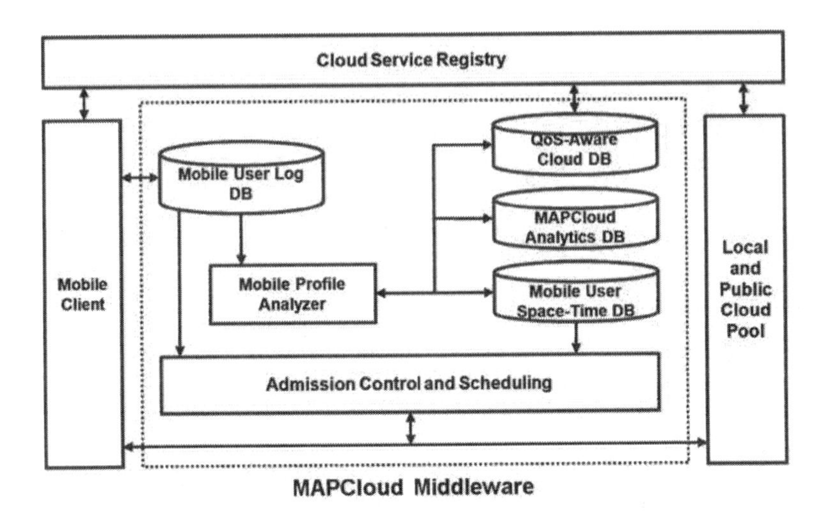

Figure 9.2: MAPCloud middleware architecture.

for Mobile Applications). CRAM combines simulated annealing and a greedy approach to optimally assign services to users.

It is worth noting that Carmen [212] provides seamless network connectivities for a set of smart devices across heterogeneous access networks, and it aims to help individual users in their own Mobile Personal Grid (MPG). Most importantly, it demonstrates the strength of using network visualization in constructing and slicing the heterogeneous network resources for different `Device Cloud`.

Instead of assuming a static network connection, ENDA [224] takes the network inconsistency into consideration. ENDA is a three-tier architecture that makes adaptable energy efficient offloading decisions in a constantly changing environment. The user mobility, server workloads and network performance are considered during the decision-making process. Therefore, it can make more efficient offloading decisions than existing approaches that neglect the issue of network inconsistency. ENDA considers not only the network heterogeneity, but also the dynamics upon it.

In [225], the authors propose an intelligent network access strategy for pervasive users to meet the user required QoS. They have used the *Intelligent Radio Network Access* (*IRNA*). IRNA is an efficient model for dealing with the dynamic and heterogeneities of multi-networks and services in MCC. Figure 9.3 shows the context management in IRNA. This architecture has three main components: *context provider*, *context broker*, and *context consumer*. Context provider gathers data from different resources in the MCC environment such as wireless network connectivity and Web services. Context broker could provide the required services for context consumers through search and lookup of

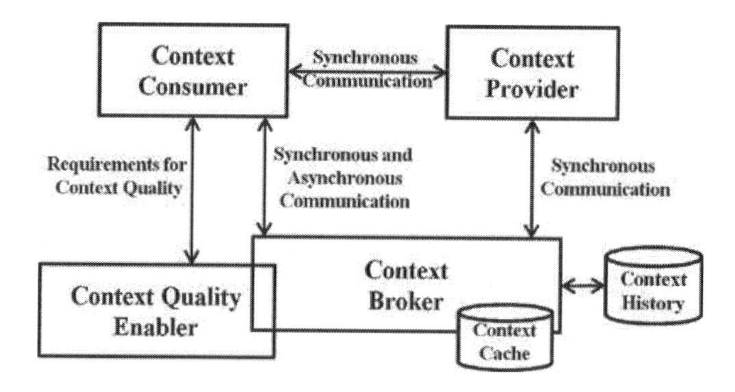

Figure 9.3: Context management architecture in IRNA.

context providers. In this architecture, when a context consumer wants to communicate with a context provider, the context consumer will request the URI (Uniform Resource Identifier) of context providers at the context broker. Using this URI, the context consumer can communicate directly with the context provider and request the context data.

9.3.3 *Enhancing Security and Protecting Privacy*

Security and privacy protection are the two main and challenging issues in CPS sensory environments, especially when powered by MCC like private and public clouds for m-commerce, m-health, and mobile social network-based CPS applications and services on-the-move. They add a high degree of system complexity to CPS applications which requires intensive computation and storage. By providing scalable and elastic resources, cloud computing could easily tackle these problems.

Because MCC utilizes the resources on the cloud to extend the capacity of a smart device, mobile cloud security naturally inherits the security concerns from both *cloud computing* [226] and *mobile security* [227]. However, the adaptation process of security techniques from traditional studies should be justified with the cases of not only limited local resources but also more potential threats, since MCC includes a vastly divergent, software, hardware, operating system, and network connection.

Most of the recently proposed mobile cloud security frameworks tend to offloads security/privacy related tasks to the cloud. [228]. Although this offloads the most of the security overhead to the cloud, mobile devices still need to perform security operations, such as authentication and encryption before outsourcing the sensitive data or tasks. These security operations still consume a considerable amount of energy and cause performance degradation.

However, this issue can be tackled by reducing the amount of data that needs to be processed [229], and optimizing the cryptographic techniques [230].

Security concerns both users and service providers because a third party may take advantage of confidential data. In addition to providing traditional computation services, MobiCloud [206] also enhances the security operation of the ad hoc network itself. As a typical example of handling security concerns in `Device Cloud`, MobiCloud proposed the service called Virtual Trusted and Provisioning Domain (VTaPD) which essentially uses programmable routing to handle information flow in various security domains.

In [231] the authors define two levels of security services; *critical security* (CS) and *normal security* (NS). CS provides strong security protection such as using longer key size, strict security access policies, isolations for protecting data. The CS service usually occupies more cloud computing resources and has benefits for cloud providers. The authors try to optimally use these security services to increase the benefits of cloud resource providers. They proposed a model called *Security Service Admission Model (SSAM)* based on *Semi-Markov Decision Process (SMDP)* to model the system reward for the cloud providers. SMDP provides a framework when the outcome of the decision making results are partly random and partly under the control of a decision makers. It is a discrete time stochastic control process where, in each state, the decision maker may choose any action that is available in state. It will react at the next time step by randomly changing state into a new state, and giving the decision maker a reward.

Privacy is an issue because the user's personal program and data are in the cloud that is not controlled by the user. In using the mobile service scenario, especially in location based services (LBS), mobile users query a server for nearby points, but they don't want to disclose their location information [232]. In general there are two main techniques for privacy, *cloaking* and *domain transformation*. As shown in Figure 9.4 (a) in spatial cloaking exact user location q will be expanded and cloaked into a region Q_0. That will make it impossible for a server to reconstruct q from the region Q_0. The server will make some candidate points of interest (PIOs) for each point in Q_0. This result will then subsequently be refined by the users trusted mobile client. Figure 9.4 (a) shows the case where Q_0 is a rectangle. In this case, the LBS server reports the POIs as $p_0, p_1, ..., p_n$. While the mobile user could be at any possible location in Q_0, all POIs inside Q_0 will be reported . Because of the large number of POIs, the cloaked query will have huge processing and communication costs.

In addition to the conventional "cloaking" techniques, the location based fined-grained access control (LFAC) [233] novelly separates users' identities from their location information. The authors aim to construct the LFAC framework to provide access control and user privacy protection for LBS. In addition, LFAC facilitates fine-grained access control and authentication by adapting a novel cryptographic access control framework that is based on a spatiotemporal predicate-based encryption.

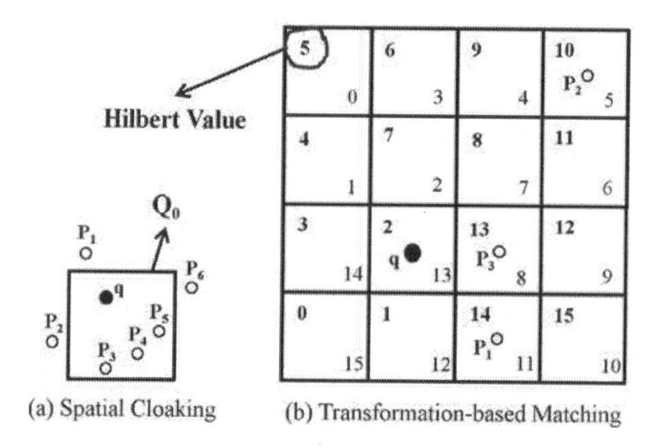

Figure 9.4: Different location-based privacy techniques.

In transformation-based matching technique (see Figure 9.4(b)), the query is evaluated in a transformed space, in which the *points and/or distances between points are encoded*. The *weak point* here is that the query result may not be accurate. To address this problem one approach defines a specific *Hilbert* ordering with a *key H*, whose value is known only by the client and a trusted broker. Not having the key value, the server cannot decode a Hilbert value into a point. In preparation for querying, the trusted entity transforms each POI into Hilbert value $\mathcal{H}(p_i)$ and it will be uploaded to the server. At query time, the client q submits its Hilbert value $\mathcal{H}(q) = 2$ to the server. The server then reports the closest Hilbert value $\mathcal{H}(p_2) = 10$ of $\mathcal{H}(q)$, which is eventually decoded by the client into point p_2.

9.3.4 Economic and Business Model

So far this chapter has been all about performance and energy saving. In this section we shall discuss the business model that can distribute the profits fairly among the players.

To improve the resource utilization and revenue, mobile service providers can cooperate to form a coalition and create a resource pool for users running mobile applications [234]. The admission control of this cooperative environment has been developed based on optimization formulation. Also revenue sharing among cooperative providers has been introduced based on a coalitional game (i.e., linear programming game).

Economic and business aspects are the new and hot open areas in cloud computing and MCC [235]. MCC contains interactions of many different businesses like mobile and cloud service providers. All of them have different criteria and policies for service providing and customer management and interaction. This will open a new direction on how to orchestrate these elements efficiently.

There are some solutions in MCC literature for efficient service management. The authors of [235] proposed a service decision making system for inter-domain service transfer to balance the computation loads among multiple cloud domains in MCC. They focused on maximizing the rewards for both the cloud system and the users, by minimizing the number of service rejections which degrade the user satisfaction level. To this end, they formulated the service request decision making process as a SMDP. In this work the optimal service transfer decisions are obtained by jointly considering the system incomes and expenses. They showed that the SMDP system can significantly improve the system rewards and decrease service disruptions.

Cloud federation [236] proposed to solve the unified business model for cloud provider interactions. It provides a flexible collaboration of cloud providers to lower the effect of *vendor lock-in* and *underestimate* problems. Vendor lock-in is a situation where cost of switching from one cloud provider to an other surpasses the benefits of the customers. The lack of *standardized service interface, protocols, and data format* is the potential threat of vendor lock-in. This will lead to underestimated problems in investment on cloud computing services. Cloud federation was proposed to solve this problem. It provides a unified, robust, and flexible cloud marketplace in which all service providers and customers could collaborate and do business easily. It has the following main components:

- *Provisioning Engine*: It maps and negotiates to allocate resources for customer applications according to market rules, policies, and standards.
- *Distribution Manager*: It guarantees the QoS and requirements of customers according to market policies.
- *Resource Manager*: It manages all resources in a unified way or provides the standard interface for different cloud platforms and customers.

9.4 Future Directions

In this section, we present future research directions within the context of task offloading, heterogeneity, security and privacy, and economic and business models.

9.4.1 Managing the Task Offloading

9.4.1.1 Scalability in the Device Cloud

In the mentioned frameworks, the authors did not show any deep studies of *scalability*, *elasticity*, and *user mobility* in the performance of their architectures when there is a large number of users (i.e., smart devices) using cloud to provide CPS application services. While resources on cloud scale up well, the communication bandwidth does not have this property. For example the performance of WiFi will drop very fast when the number of users is high. There should be a study which considers the effect of aforementioned factors on the system performance.

9.4.1.2 Making the Offloading Decision Process Transparent to the Application Developer

While the up-running CPS applications are treated as a collection of reusable components, the offloading decision-making process should be designed and developed accordingly to different requirements for various CPS applications and services. The system can thus provide highly customized offloading policy groups from which a CPS application developer can select the suitable one according to the features of application alone, e.g., how heavy does each component rely on computation and real-time information from the cloud. The parameters from cloud environment should not be the concern of the developer, at least not on the level of each component of the application.

9.4.1.3 Context Awareness on Trading Off the Optimization between Performance Improvement and Energy Saving

Both power and execution time efficiency of offloading are heavily studied in the MCC literature. However, there is a lack of research that aims to simultaneously optimize the power consumption and execution time. We reckon it as a promising research direction because a real-world CPS application may have different degrees of QoS requirements among these two objectives, and more importantly, this demand may vary under different circumstances. For example, on a fast moving vehicle under complex traffic conditions, the GPS device demands real-time calculation but should not have to worry about battery consumption. However, if this GPS device is unplugged by the driver to complete the rest of journey on feet or bike, the demands among the two objectives should be able to adapt automatically. *Multi-objective optimization techniques* appear to be a natural solution to satisfy this requirement by providing adapting configuration to the smart device. This advice concerns the research on all three types of MCC-enabled CPS architectures.

9.4.1.4 Tasks Distributing among Sensors

CPS applications and services are mainly sensory-based which are producing streams of data like traffic monitoring services. They need real-time processing and huge data storage. Future research directions should pay more attention to how to distribute the processing and storage tasks among sensors and cloud to optimize power usage, execution time, and paid price. After all, study on balancing the workload within a `Device Cloud` is still rare in the field of MCC to support diverse CPS applications.

9.4.1.5 Offloading Decision Making in a `Hybrid Cloud`

To the best of our knowledge, most of the existing studies on offloading techniques are focusing on either the decision on offloading tasks to the `Internet Cloud` or within the `Device Cloud`. However, we reckon the `Hybrid Cloud` should be the emerging direction for future research. Because a CPS sensory environment cannot fully depend on one kind of offloading technique, because the characteristics of the resources on the Internet or peer smart devices are fundamentally different, as shown in Figure 9.1, and the cost of utilizing the available resources from different sources also varies. Thus, the offloading decision in a hybrid CPS sensory environment cannot be made by easily combining the existing techniques. Therefore, studying the various characteristics of offloading in a `Hybrid Cloud` architecture is the key to maximizing the leverage provided by both `Internet Cloud` and `Device Cloud`.

9.4.2 Encountering Heterogeneity

Although some middleware solutions have been proposed for MCC, this area still has many open questions that should be explored.

9.4.2.1 Efficient Middleware

The main source of the Web traffic on the Internet these days is related to multimedia-based data. According to an eMarketer prediction, mobile music, multimedia sharing, and game users will increase by 55% from 2009 to 2014. This will open a new research direction which considers efficient middleware frameworks for achieving high QoS in a CPS sensory environment for the aforementioned multimedia-based CPS applications [237, 238]. It is clear that efficient middleware will be applicable to `Internet Cloud`-based CPS architecture.

9.4.2.2 Dynamic Adaptive Automated System

To add on the complexity of the heterogeneities, the CPS sensory environment is rapidly changing during the deployment of any system that was manually configured by an engineer. The adaptivity of the system is at only a glacial pace. This fact forces highly skilled engineers to waste significant time on manually implementing the detailed configurations. Often, the result is still equally inflexible, forcing users to also rely on their innate human adaptivity to find "workarounds." The heterogeneity of the environment naturally provides a great deal of flexibility, but we reckon that it is promising to enable the system to use computational search to achieve its full potential for flexibility and adaptivity. For example, heuristics is one of the most promising techniques to enable automation and dynamic adaptation of the system. We should be creating new ways to develop, deploy, and maintain CPS systems. Apparently, a dynamic adaptive automated system is the ultimate solution to ease the issue of heterogeneity in a mobile cloud network, especially in a `Hybrid Cloud` network.

9.4.2.3 Mobile Big Data

Different CPS stakeholders are producing huge data, such as location-time service usage logs, mobile data traffic logs, cloud resources usage logs, wired and wireless network usage logs, etc. The study of this "*Mobile Big Data*" (for example using data mining techniques) will open a new research direction in the CPS middleware design to efficiently and optimally manage and allocate resources in CPS sensory environments to achieve high QoS.

The authors of [223] present a comprehensive survey on MCC with major focus on heterogeneity. The authors describe the challenges and opportunities imposed by heterogeneity and identify hardware, platform, feature, API, and network as the roots of MCC heterogeneity. They explain major heterogeneity handling approaches, particularly virtualization, service oriented architecture, and semantic technology. However, the computing performance, distance, elasticity, availability, reliability, and multi-tenancy of remote resources are marginally discussed in these studies which necessitates further research to explain the impact of remote resources on the augmentation process and highlight paradigm shift from the unreliable surrogates to reliable clouds.

9.4.3 Enhancing Security and Privacy

9.4.3.1 Finding Protection Solutions That Are More Efficient Is Still a Research Topic

Due to the limited resources and capabilities of portable devices and sensors in CPS ecosystems, finding solutions that are more efficient will remain the research focus in the foreseeable future. Specifically, the future research direction in designing security protocols and algorithms should focus more on power-efficiency and delay caused by such security algorithms and how to partition them on cloud and mobile devices.

9.4.3.2 Context Awareness on Dynamic Security Settings

Because of user mobility, the security level of mobile users will change from location to location; for example changing from 3G network coverage to WiFi network coverage or from local cloud resources to public cloud. There are some questions that should be explored in future research like: How could this information affect the design of security policies and algorithms for mobile users? What are the effects of using different security policies on power consumption, delay, and application performance?

9.4.3.3 Trade Off between the Functional Performance Degradation and Security and Privacy Requirements

The operations taken to provide insurance on security and privacy consume energy and cause delay. Due to the fact that security and privacy are essentially *non-functional requirements*, in addition to "do it smartly" by enhancing the efficiency of existing precaution techniques, more attention is demanded on the study of relaxing the security and privacy requirements. For example, if the loss caused by potential threat can be measured, insured, and reimbursed by a third party, user and CPS service provider should both have the ability to downgrade to a range of relaxed precaution settings for the purpose of ensuring a specific level of functional performance. This "doing less" approach introduces the risk that is affordable to the individual user and the service provider.

9.4.4 Economic and Business Models

Cloud federation concept could be extended to a "*Mobile Cloud Federation*" concept which opens a new research direction to have unified and robust CPS cloud environment. It will open some important questions that should be answered in future research:

- From a *game theory perspective*: how could we design policies and mechanisms to have stable a "Mobile Cloud Federation"? Does this market have "Nash Equilibrium"?
- With different players in MCC-enabled CPS sensory environments, what are the best strategies to optimize cloud providers' revenue while having *green cloud computing*? What are the best strategies for maximizing the user's saving on price, delay, and power consumption?
- What are the best strategies and policies to offer competitive prices to mobile users in MCC-enabled CPS sensory environments?
- As has been mentioned, one of the most important area is m-commerce applications, such as acquiring user profile location and priorities, mobile user forwarding strategy, advertisement strategy to maximum cloud providers/business owners' profit, etc.
- Service Oriented Computing (SOC) tries to provide a unified protocols and standard interface for cloud service interactions [236] like SOAP (Simple Object Access Protocol), UDDI (Universal Description Discovery and Integration), and WSDL (Web Services Description Language). This will open a new research direction to extend the SOC framework to cover other MCC player criteria like wireless network providers, mobile users and sensors.

9.5 Summary

With the immense and complex information perceiving and processing capability, MCC technologies are emerging to empower diverse CPS applications by providing ubiquitous and rich functionalities, regardless of the resource limitations of smart devices. In this chapter, we first clearly defined a novel classification methodology on three different CPS architectures: `Internet Cloud`, `Device Cloud`, and `Hybrid Cloud`. The user of `Internet Cloud` benefits from the resources that are on the Internet, while the user of `Device Cloud` benefits from the resources that come from the peer devices in the local, ad hoc, dynamic connections. We find that most existing literature on MCC is focusing only on the `Internet Cloud`, and reckon that the study of `Device Cloud` and `Hybrid Cloud` should be promoted to leverage the resource sharing among peer devices. Then, the most recent advances of the CPS applications using MCC technologies are reviewed accordingly, and challenges and opportunities are summarized and discussed. Furthermore, open research issues on other aspects of MCC, but highly relevant to CPS, such as offloading techniques, heterogeneity of the infrastructure, security and privacy, and economic/business model, have also been discussed within the context of existing schemes.

CPS APPLICATIONS

In this part, we introduce some of the potential applications where such platforms can be used in the future.

Smart Agriculture: This is an outdoor domain where sensors need to be deployed close to the environment with limited energy. Further, sensors need to cover large geographical area in comparison to indoor domains. Sensors with low cost are ideal to be deployed in large numbers in agriculture domains. Farmers can deploy sensing middleware platforms permanently. Alternatively, they can carry their mobile phones or tablets as data collection devices if the geographical area is quite small. Alternatively, sensing middleware platforms can be fixed into vehicles or robots that go around the field and collect data from sensors. These two options are illustrated in Figure III.A.

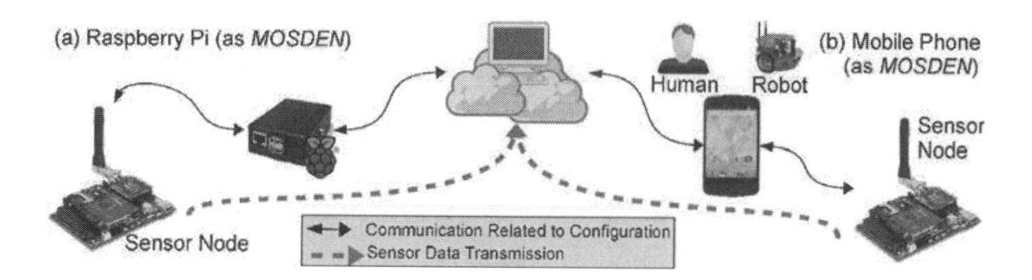

Figure III.A: MOSDEN can be used as an intermediary device where it retrieves data from low-cost sensors and pushes it to the cloud in real-time or on demand. These sensing middleware platforms can be installed on both static (e.g., Raspberry Pi) and mobile hardware devices (e.g., tablets and smartphones).

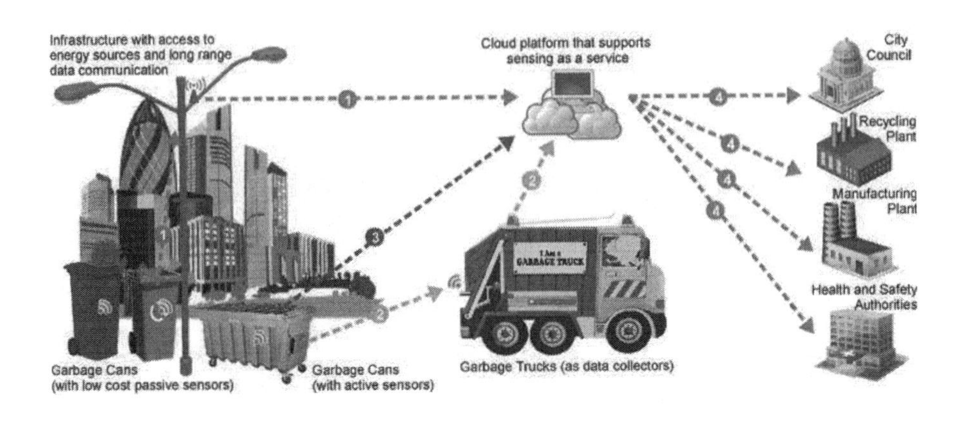

Figure III.B: Efficient waste management in smart cities supported by the sensing as a service model

Smart Home and Office: This is an indoor domain where sensors have comparatively sophisticated access to permanent energy sources. Sensor data can be collected through crowd sensing [239] techniques where corporate smartphones will play the role of intermediary. In comparison to other domains, the smart home and office domain is highly dynamic. Internet Connected Objects (ICO) may appear, disappear, and move faster than in any other domain. Additionally, short range protocols may work in most cases. It is also important to highlight that real-time data collection is critical in this domain. Further, event detection and actuation is an integral part of this domain where occupants receive increased quality of life or workspace.

Smart Cities: Mobile and static sensing middleware platforms are useful in a wide range of sensing tasks that need to be carried out in the smart city domain. One such example is waste management. Sensors are expected to be fitted into garbage cans deployed around the modern cities. These sensors will collect information about the content of the garbage, garbage level, cleaning and pick times and dates and so on. The mobile sensing platforms can be fitted into garbage collection trucks, and other municipal council owned vehicles. Such data received on time has a significant value to many different parties. Static sensing platforms can be fitted into nearby public infrastructure such as light poles. Such infrastructure has permanent connectivity to power grids and can support long-range 3G communication. For example, city councils may use sensor data to develop optimized garbage collection strategies, so they can save fuel costs related to garbage trucks. Additionally, recycling companies can use sensor data to predict and track the amount of waste coming into their plants. This allows them to optimize their internal processes. Further, health and safety authorities can monitor and supervise the waste management process without spending a substantial amount of money for manual monitoring inspections. The scenario of waste management is illustrated in Figure III.B.

Chapter 10

Connected Healthcare for CPS

Chi Harold Liu

Beijing Institute of Technology, China

CONTENTS

The preceding chapters discuss the enabling technologies of CPS, including the aspects of fundamental networking technologies, M2M communications, and mobile cloud computing. In the following chapters, we will consider the enormous applications of CPS. We start with the applications of CPS technology in connected healthcare in the present chapter.

10.1 Introduction

One of the major challenges around the world recently has been the continuous increase of the elderly population, and thus the delivery of quality care while reducing healthcare costs is highly needed [240]. Particularly with the continuing advances in sensors and sensor-supporting technologies including pervasive computing and communications capabilities, we are witnessing an emergence of a variety of promising applications stemming from the integration of sensing and consumer electronics, allowing people to be constantly monitored [241].

Family-based healthcare services [242, 243] render the patient's full freedom at home, which dramatically reduces the need and waiting time for face-to-face contact with care professionals, where the healthcare providers remotely monitor the patient's physical condition 24/7, even when the patient is mobile. Furthermore, if the measurements show certain deterioration of the patient's well-being, alerts are generated and sent to the patient's mobile phone. The care professional's help could also be invoked immediately as part of the service requirement. Not only the elderly and chronically ill but also working parents may derive benefits from these systems for delivering high-quality care services for their babies and little children. Also, the benefits can be extended to young fitness trainers who are interested in continuous monitoring of their training outcomes. Hence, a family-based system should provide *collaborative*, *interactive*, and *long-term* support to all users, based on the powerful data processing and analysis units, but such a system barely exists within the research community.

In this chapter, we present "HealthKiosk," a family-based healthcare monitoring system that bridges the data centers and biomedical sensors. As shown in Figure 10.1, biomedical sensors collect personal data on health conditions and other vital signs, and report them to the sensor proxy, where the latter serves as a bridge between the sensor network (e.g., via WiFi, Bluetooth, RS232) and the kiosk controller. We note that the sensor proxy will also maintain some data processing logic, like correlating the blood pressure data with the patient's social security card meta data. Then, the developed kiosk system automatically connects to a variety of backend servers like a clinical

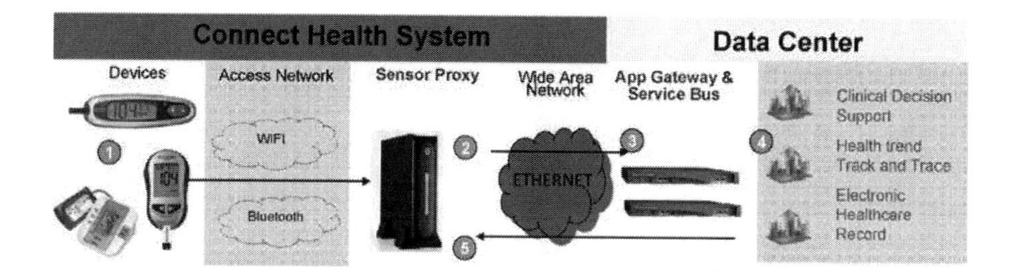

Figure 10.1: A system diagram for the proposed HealthKiosk system, where (1) biomedical sensors send the sensor readings to the sensor proxy, (2) the sensor readings are processed and sent to the kiosk via Ethernet, (3) and (4) the data are further analyzed by the backend server, and (5) treatment suggestions are sent to the sensor proxy.

supporting system in major hospitals via a wide area network (WAN, e.g., the Ethernet) for further treatment. For instance, after receiving the personal medical data, the application may decide to remeasure the blood pressure, and the sensor proxy will command the sensor accordingly.

We believe that the proposed solution provides a universal research asset to enhance the clinical process for streamlining interactions among care professionals and patients. It is also worth noting that the HealthKiosk system has been piloted and deployed at the Peking University People's Hospital (PKUPH[1]) since February 2010, and has shown its success so far.

The rest of the chapter is organized as follows. In Section 10.2, we highlight related research activities. Section 10.3 establishes a formal model of our system. Section 10.4 describes the design of the sensor proxy and related technical challenges and solutions, and Section 10.5 presents the HTTP interface used between the kiosk controller and the sensor proxy. Two complete case studies are presented in Section 10.6, followed by the conclusions and future work in Section 10.7.

10.2 Related Work

Healthcare systems using either mobile or other equipment have been proposed and studied for over a decade to provide convenient and constant monitoring of a patient's health conditions [244, 245, 246, 247]. It has been identified as an important branch of research of the machine-to-machine (M2M [248]) or, the Internet of Things (IoT [249, 250]), that aims to connect physical sensors to the Internet for better monitoring and analysis. Recently, low-cost and

[1]PKUPH, is one of the biggest hospitals in China, founded in 1918, housing more than 2,400 staff and admitting more than 40,000 inpatients a year. It receives more than 1,560,000 outpatients, and more than 133,000 emergency cases annually.

effective sensors, such as blood pressure and heart rate sensors [251], have been made available to a large number of individuals and families for healthcare solutions.

There are several case studies that take advantage of the above development to offer healthcare solutions to particular groups of patients in limited regions. For example, Columbia University monitors thousands of diabetes patients in the greater New York area [252]. There are also evaluations and actual field deployment to confirm that mobile Internet devices such as cell phones can be used as gateways to connect body area networks to the Internet [253, 254, 255]. Furthermore, efforts have been taken to define standards for mobile healthcare systems [256].

In this chapter, we emphasize our architecture and framework, which can accommodate various sensors, a wide range of Internet connectivities, and comprehensive sharing and analysis of the data. We provide a middleware appliance, named "sensor proxy" that can speak the languages of both the physical world and the Internet, to bridge the gap and mediate the interactions between them. It can be deployed as a stand-alone box or part of broadband modems/routers and mobile handsets. More importantly, we design its interface to be RESTful [257, 258] such that its data can be obtained and shared to multiple parties easily. This reduces the effort and power requirements on sensors since they do not have to provide the interfaces [259], and consequently makes them even more portable and energy efficient.

10.3 System Model

The overall architectural view of our proposed HealthKiosk system is shown in Figure 10.1, where patients can use their installed biomedical sensors (e.g., glucose-meter, blood pressure monitor, etc.) to take measurements. Then, the data are sent to the sensor proxy via existing communication networks like 3G, WiFi, and Bluetooth. The sensor proxy can be a small server or even an application in the mobile handset, and it holds certain data processing logic to integrate multiple pieces of the raw data from a variety of the physical sensors to a common format prepared for uploading to the application gateway (or the "kiosk controller" in our design). Then, either the transformed and integrated data are relayed to the healthcare service supporting system in the backend (if needed), or they are locally consumed/shown on an easy-to-use touchscreen user interface (UI) at the kiosk. We also allow interaction between the backend supporting system and the patients. If his/her medical data shows the deterioration of his/her health condition detected by the clinical decision supporting system, alerts are generated and sent to the sensor proxy, like the patient's mobile phone. If necessary, the care professional's intervention could also be invoked immediately as part of the service requirement. We next describe the functional designs and the challenges of the sensor proxy in Section 10.4.

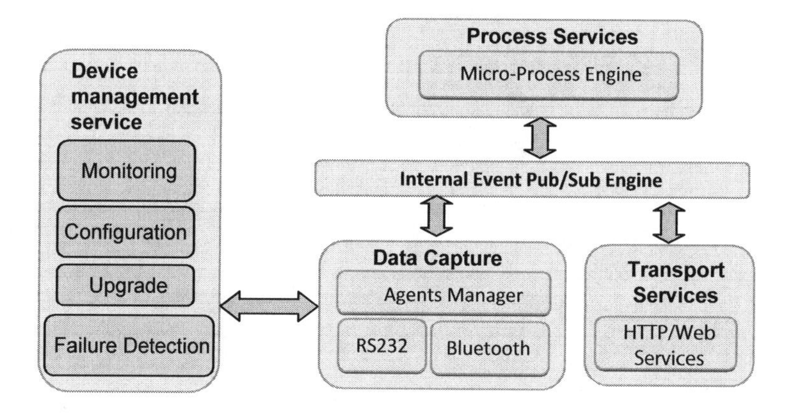

Figure 10.2: A functional design architecture for the sensor proxy.

10.4 Sensor Proxy Design

With the proposed HealthKiosk system, the sensor proxy serves as an important design element bridging different biomedical sensors with the kiosk controller, either wirelessly or wire connected. Furthermore, it also provides a certain degree of sensing event processing and sensor management functionalities. A functional design architecture is shown in Figure 10.2, and is composed of five core elements:

10.4.1 Data Capture Module

It is a wrapper with which different agents connect to the biomedical sensors via underlying communication channels (e.g., RS232, Bluetooth). An "agent manager" is implemented to load these agents and transfer the data from medical sensors to a common business event of the sensor proxy.

10.4.2 Internal Event Pub/Sub Engine

Due to the instability of the wireless conditions between the sensor proxy and biomedical sensors, the data uploads might be blocked or delayed from the sensor agents. From the user experience perspective, they may feel uncomfortable being kept waiting for the response from the sensors. Therefore, we developed a lightweight "internal pub/sub engine" (see Figure 10.2), to decouple the data from multiple physical sensors. In our design approach, the process service module subscribes the medical data of interest, and the sensor agents publish the medical data whenever it arrives. Upon receiving an event

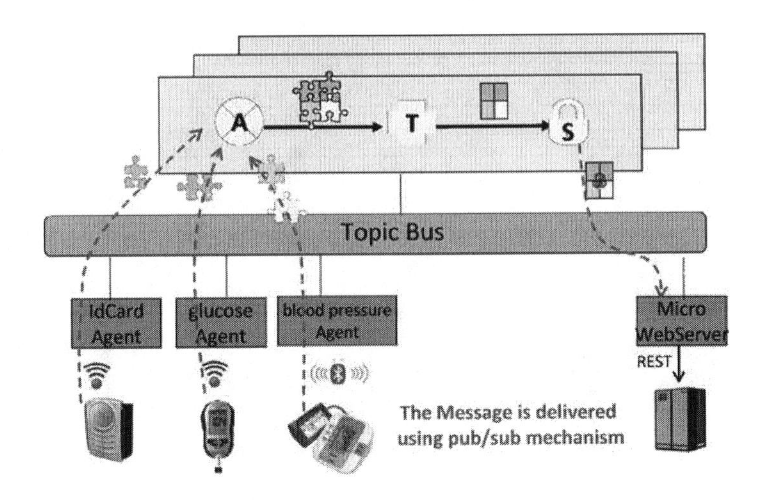

Figure 10.3: The illustrative functional diagram for data aggregation, transformation, and encryption.

from the data capture module, the internal event pub/sub engine will dispatch it into the process service module for further processing. After the processing is completed and the data is sent back to the internal pub/sub engine, the latter will choose a transport service (e.g., HTTP) to publish the event to the kiosk controller. Therefore, the user experience is significantly enhanced by our design.

10.4.3 Process Service Module

It provides a micro engine to handle the event processing step by step, for example, a typical community healthcare service will require the sensor proxy to correlate the patient's meta data from his/her social security card with the patient's sensor data, and then send the combined data through an XML format to the remote repository (like the kiosk database). Therefore, a reusable data processing and transformation capability must be provided to convert a variety of data formats to a common interface so that effective integrations among domains can be performed.

To achieve the design goal, we developed a lightweight process engine, called the "micro process engine" (see Figure 10.2), which runs a simple data processing logic, and defines some reusable data operations. As shown in Figure 10.3, the processing logic is composed of an aggregation operation ("A"), a transform operation ("T"), and a security operation ("S"). When the logic receives any data, it first aggregates them together (by using the "A" operation) and then calls for a "T" operation to transform the combined data in a certain format and use the security operation ("S") for encryptions. Figure 10.4 shows

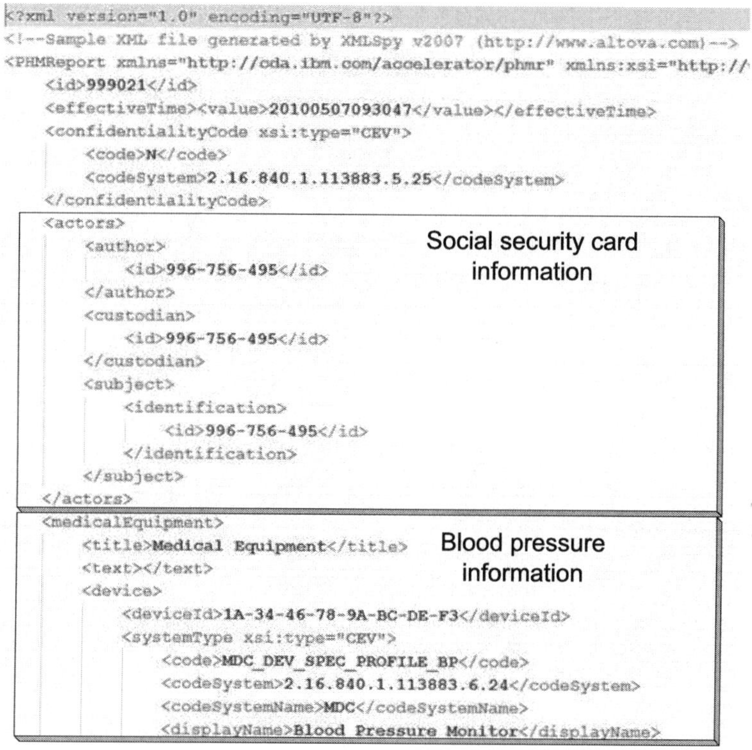

```
<?xml version="1.0" encoding="UTF-8"?>
<!--Sample XML file generated by XMLSpy v2007 (http://www.altova.com)-->
<PHMReport xmlns="http://oda.ibm.com/accelerator/phmr" xmlns:xsi="http://
    <id>999021</id>
    <effectiveTime><value>20100507093047</value></effectiveTime>
    <confidentialityCode xsi:type="CEV">
        <code>N</code>
        <codeSystem>2.16.840.1.113883.5.25</codeSystem>
    </confidentialityCode>
    <actors>
        <author>
            <id>996-756-495</id>
        </author>
        <custodian>
            <id>996-756-495</id>
        </custodian>
        <subject>
            <identification>
                <id>996-756-495</id>
            </identification>
        </subject>
    </actors>
    <medicalEquipment>
        <title>Medical Equipment</title>
        <text></text>
        <device>
            <deviceId>1A-34-46-78-9A-BC-DE-F3</deviceId>
            <systemType xsi:type="CEV">
                <code>MDC_DEV_SPEC_PROFILE_BP</code>
                <codeSystem>2.16.840.1.113883.6.24</codeSystem>
                <codeSystemName>MDC</codeSystemName>
                <displayName>Blood Pressure Monitor</displayName>
```

Social security card information

Blood pressure information

Figure 10.4: The combined data with social security card and blood pressure sensor information in one XML data format.

a snapshot of the real XML code obtained after combining the social security card data and the blood pressure measurement.

10.4.4 Transportation Service Module

While the data capture module connects the sensor proxy to the sensor network, the transportation service module connects the sensor proxy to the Internet, and ultimately the kiosk controller. We leverage the RESTful design approach [257, 258], where the sensor proxy behaves as a HTTP server (while leaving other methods, e.g., MQTT [260], for future extensions), and receives the instructions from the kiosk server by HTTP request and methods (like `GET` and `POST`). In the meantime, the sensor proxy also acts as an HTTP client and uploads the measured data and its associated patient's meta data to the kiosk controller database where the relational database is stored. The detailed HTTP interface from the kiosk server to the sensor proxy and backwards is described in Section 10.5. Besides the fundamental capability of delivering the data back to the backend server, it is worth noting that the sensor proxy itself can also have certain exposure for the medical data in a Web-friendly manner, and thus the application developer can "mash-up" these data and show them in the mobile phone or desktop for patients.

10.4.5 Device Management Service Module

It provides a common infrastructure to configure, deploy, monitor, and update other modules. In different cases, the sensor proxy might connect to different medical sensors, transform the sensor data into different formats, and expose the data in different ways. Therefore, the device management service will configure and deploy different agents in data capture modules, and different processes in the micro process engine for these cases.

10.5 HTTP Interface

We next describe a few important HTTP interfaces connecting the kiosk controller and the sensor proxy, including naming and addressing, initializing and obtaining sensor readings.

10.5.1 Get Naming and Addressing

We use the HTTP `GET` method by calling the URL `http://hostname:port/healthcare/namingaddressing`, to obtain the naming and addressing information of the sensory proxy and its connected biomedical sensors. Since multiple physical biomedical sensors connect to the sensor proxy simultaneously, we need an efficient naming and addressing mechanism so that the sensor readings can be identified, and the control commands can reach the physical world. We leverage the RESTful style [257, 258] as:

10.5.1.1 *Sensor Proxy Naming*

We name the sensor proxy 1 as:

```
<sensor_proxy name="proxy_1">
```

10.5.1.2 *Biomedical Sensors Naming*

For instance, the following pseudocode shows the naming mechanism of a social security card reader connected with the sensor proxy 1.

```
<sensor_proxy name="proxy_1">
<sensor type="social_security_card">
```

10.5.1.3 Biomedical Sensors Addressing

We use a URL-like address to hierarchically differentiate multiple sensors as:

```
<sensor type="bloodpressure">
<start_address>
  http://hostname:port/healthcare/
     bloodpressure/
     1A-34-46-78-9A-BC-DE-F3/start
</start_address>
```

10.5.2 Start Blood Pressure/Glucose Reader

We use the HTTP `POST` method by calling `http://hostname:port/healthcare/glucose/deviceID/start`, to start the sensor reader, e.g., deviceID=1A-34-46-78-9A-BC-DE-F3 uniquely denotes the blood pressure sensor. The request complies with the RESTful [257, 258] style, and the following returns may be generated:

(1) 200: OK, if the sensor is successfully started.

(2) 404: Not found, if one cannot find the corresponding sensor.

(3) 400: Bad request, if other errors in the request format exist.

(4) 500: Internal server error, identified in the sensor proxy.

10.5.3 Get Social Security Card ID

We use the HTTP `GET` method by calling `http://hostname:port/healthcare/socialsecuritycard/1A-00-00-00-00-00-00-01/data`, to obtain the social security card ID from the kiosk controller, once the data is available. If not, the sensor proxy will return `200` but with an empty response body; nevertheless, for the successful data fetching, a `200` return will be generated with the body of <id hasCard="true">996-756-495</id>. The sensor proxy can also return <id hasCard="false"/>, which indicates that there is no card in the card reader. The following items summarize the possible returns:

(1) 200: OK with body either <id hasCard="true">996-756-495</id>, or <id hasCard="false"/>, or empty.

(2) 404: Not found, if one cannot find the corresponding device.

(3) 400: Bad request, if there are other errors in the request format.

(4) 500: Internal server error, identified in the sensor proxy.

10.5.4 Get Blood Pressure/Glucose Data

We use the HTTP `GET` method by calling `http://hostname:port/healthcare/bloodpressure/deviceID/data`, to obtain the data from blood pressure and glucose sensors. Based on the patient's current vital signs, the kiosk controller tries to get blood pressure data from the sensor proxy, where the message is flowing from the kiosk controller to the sensor proxy. The obtained measurements use an XML format and are embedded in the body of the HTTP response. The following items summarize the returns:

(1) 200: OK with empty body.

(2) 404: Not found, if one cannot find the corresponding device.

(3) 400: Bad request, if there are other errors in the request format.

(4) 500: Internal server error, identified in the sensor proxy.

(5) 200: OK with body of the combined XML data.

10.6 Case Studies

To better illustrate the rich set of functionalities our system is able to provide, we present two complete case studies in this section: one is the stationary HealthKiosk system where a PC-like sensory proxy is developed to connect the biomedical sensors (see Figure 10.5 and Figure 10.6), while the other one is to use the mobile handsets behaving as the sensor proxy, or the "mobile HealthKiosk," for the support of mobile healthcare solutions (see Figure 10.7).

10.6.1 Stationary HealthKiosk

Figure 10.5(a) demonstrates an implementation of the HealthKiosk system, composed of a kiosk touch-screen UI, a social security car reader to leverage the meta data of the patients, a height and weight scaler, and a blood pressure sensor. Figure 10.5(b) shows that a colleague of IBM Research – China is taking a blood pressure measurement with his social security card inserted into the card reader. The detailed touch-screen UI is demonstrated in Figure 10.6, where Figure 10.6(a) shows the welcome page, and by clicking the "next" button, the user will be directed to choose taking either height and weight measurements, or blood pressure measurements. The measurements of each sensor are shown in Figure 10.6(b) and Figure 10.6(c), respectively.

It is also worth noting that our stationary HealthKiosk system has been piloted and deployed at the PKUPH for diabetes treatments by building an evidence-based clinical care solution focusing on chronic disease management. Extensive training has been offered by the IBM Research – China, and the

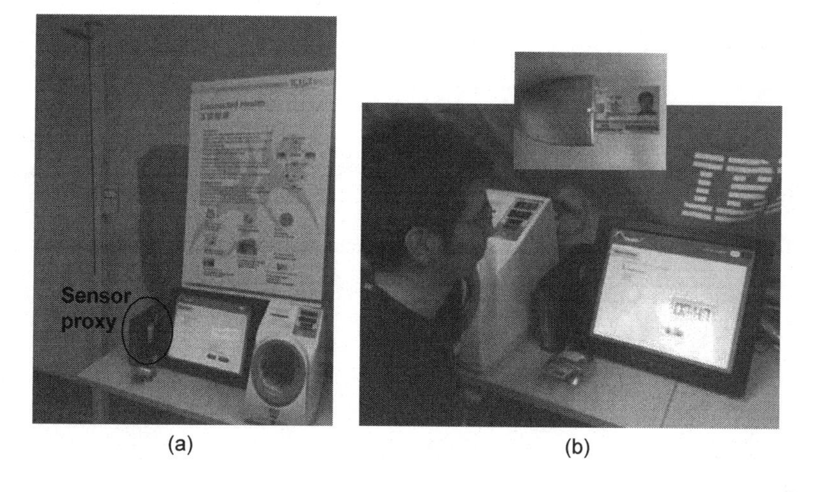

Figure 10.5: An illustrative case study showing (a) the overall system composed of a touch-screen kiosk, a sensor proxy, a social security card reader, a height and weight sensor, and a blood pressure sensor, and (b) a colleague of IBM Research – China is taking blood pressure measurement with his social security card.

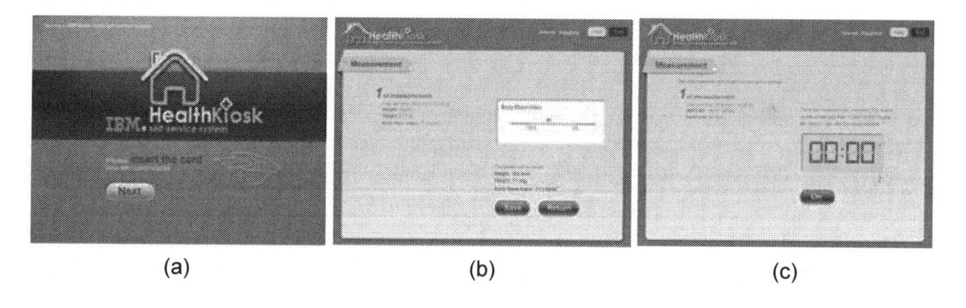

Figure 10.6: Developed UI, showing (a) the welcome page, (b) the height and weight measurement, and (c) the blood pressure measurement.

current contract will be extended to the next year, when more systems and complicated offerings will be provided. Famous Chinese newspapers including *China Daily*, Reuters, and ZDNet have covered the news.

10.6.2 Mobile HealthKiosk

To take advantage of the mushrooming popularity of the mobile devices, we also implement a *mobile* HealthKiosk system, complementing the previous

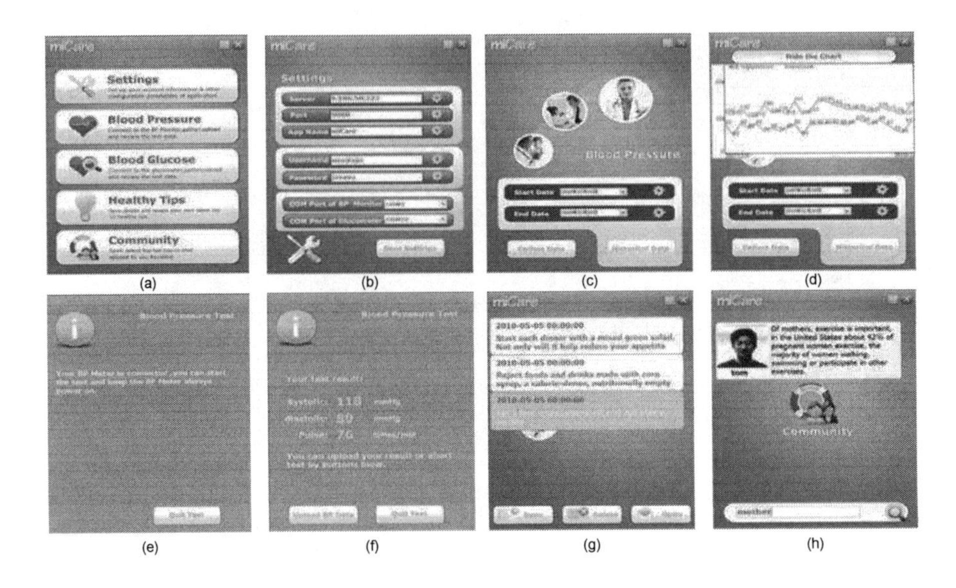

Figure 10.7: The case study for the mobile HealthKiosk system, showing (a) the welcome page, (b) the personal settings, (c) and (d) the blood pressure measurement entrance and the historical data trend, (e) and (f) new measurement, (g) and (h) healthy tips and community suggestions.

stationary sensor proxy, as shown in Figure 10.7. Different from the stationary sensor proxy, we use the mobile handset to behave as a sensor proxy; and the developed lightweight process service module periodically listens to the wireless connections of the biomedical sensors to obtain the newly arrived medical data. Then, the process engine helps with the data processing, e.g., to transform or enrich the data into a suitable format for upload. Apart from all these, further applications developed include the kiosk applications and a community widget. The kiosk applications provide an easy-to-use user interface for patients to interact with the biomedical sensors. The community widget lets the patient leverage resources from the mobile healthcare community, e.g., the care professionals and hospital equipment, where patients are able to receive good tips from other patients or doctors.

Figure 10.7(a) shows the welcome page summarizing the functionalities we provide in a user-centric and service-oriented manner, where patients are able to set their personal settings, take blood pressure and glucose readings, and browse some healthy tips and community suggestions. A typical scenario using our deployed system is described as follows. James is a fifty-year-old man, who has suffered high blood pressure for two years. Provided by our mobile healthcare system, he owns a blood pressure sensor and glucose-meter. Furthermore, his personal mobile phone has installed our developed widgets and the mobile

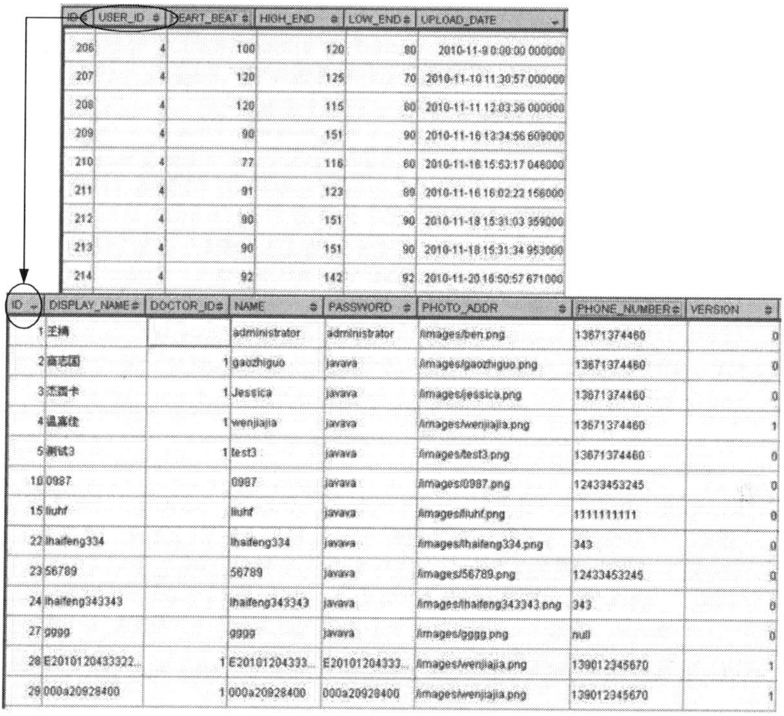

ID	USER_ID	HEART_BEAT	HIGH_END	LOW_END	UPLOAD_DATE
206	4	100	120	80	2010-11-9 0:00:00 000000
207	4	120	125	70	2010-11-10 11:30:57 000000
208	4	120	115	80	2010-11-11 12:03:36 000000
209	4	90	151	90	2010-11-16 13:34:56 609000
210	4	77	116	60	2010-11-18 15:53:17 046000
211	4	91	123	89	2010-11-16 16:02:22 156000
212	4	90	151	90	2010-11-18 15:31:03 359000
213	4	90	151	90	2010-11-18 15:31:34 953000
214	4	92	142	92	2010-11-20 18:50:57 671000

ID	DISPLAY_NAME	DOCTOR_ID	NAME	PASSWORD	PHOTO_ADDR	PHONE_NUMBER	VERSION
1	[illegible]		administrator	administrator	/images/ben.png	13671374460	0
2	[illegible]	1	gaozhiguo	javava	/images/gaozhiguo.png	13671374460	0
3	[illegible]	1	Jessica	javava	/images/jessica.png	13671374460	0
4	[illegible]	1	wenjiajia	javava	/images/wenjiajia.png	13671374460	1
5	[illegible]	1	test3	javava	/images/test3.png	13671374460	0
10	0987		0987	javava	/images/0987.png	12433453245	0
15	liuhf		liuhf	javava	/images/liuhf.png	1111111111	0
22	lhaifeng334		lhaifeng334	javava	/images/lhaifeng334.png	343	0
23	56789		56789	javava	/images/56789.png	12433453245	0
24	lhaifeng343343		lhaifeng343343	javava	/images/lhaifeng343343.png	343	0
27	gggg		gggg	javava	/images/gggg.png	null	0
28	E2010120433322...	1	E20101204333...	E20101204333...	/images/wenjiajia.png	13901234567 0	1
29	000a20928400	1	000a20928400	000a20928400	/images/wenjiajia.png	139012345670	1

Figure 10.8: Historical records of blood pressure measurements and the relational database containing the patient information.

sensor proxy application. Every day, James is able to periodically take measurements by himself; then, his mobile phone obtains the medical data from the blood pressure sensor and glucose-meter. Figure 10.7(g) shows a measurement result, i.e., 118/89mmHg for blood pressure and 76 times/min for pulse; and Figure 10.7(h) shows the tips received from the community group members. Figure 10.8 shows the relational database storing the patient's personal information and all historical measurements.

10.7 Summary

In this chapter, we presented a novel family-based healthcare monitoring system, called "HealthKiosk," with its detailed designs and case studies. The proposed solution is patient-driven and service-oriented, and provides a user-friendly interface for visible patient care. HealthKiosk could potentially minimize the efforts of care professionals, not only applicable for the elderly but also children and young fitness trainers. The proposed architecture leverages the RESTful design style, nevertheless introducing a novel development of the sensor proxy, both in the PC style and as a mobile widget. The sensor proxy behaves not only as a bridge between the raw sensor readings and the kiosk controller, but also as a data processing logic to integrate, correlate, and transform multiple pieces of data to an XML format. We demonstrated the system performance by showing two complete case studies for both stationary and mobile sensor proxies used in the proposed end-to-end HealthKiosk system. Finally, the proposed solution has been piloted and deployed at the PKUPH for diabetes patients, via building an evidence-based clinical care solution focusing on chronic disease management in China. In the future, we are planning to investigate an integrated mobile monitoring platform across multiple industry domains, like healthcare and logistics, and make it applicable in real scenario deployments.

Chapter 11

Multi-Player Gaming for Public Transport Crowd

Chi Harold Liu

Beijing Institute of Technology, China

CONTENTS

Besides healthcare applications, multi-player gaming on smart devices is catching more and more attention recently, particularly in public transport environments where people are looking forward to playing rich contextual games while traveling. In this chapter, we will discuss how to enable highly interactive and opportunistic mobile gaming in public transport environments. Our focus is a mobile gaming platform that utilizes both the cloud resource and the crowd resource.

11.1 Introduction

In many crowded metropolitan cities, people often travel by public transport like metro and local buses everyday. A finding shows that about 3,000 commuters in the west of England spend about 139 hours a year on average traveling to and from their workplace [261]. Therefore, mobile gaming, or computing in general, in crowded public transports can open up a new opportunity to harness a collection of smart devices (e.g., smartphones and tablets [262, 263]) in their vicinity as a unified computing substrate. Figure 11.1 shows a number of snapshots taken in the London Underground where a large number of passengers are looking at their smart devices, either reading news, checking emails, or playing games. A promising aspect we focus on in this chapter and our research path in its own right, is to enable *highly interactive* and *opportunistic multi-player* gaming [264, 265, 266] among the crowd, based on their travel trajectories, gaming preferences, and other related social information in different contexts. Examples of these games are, but not limited

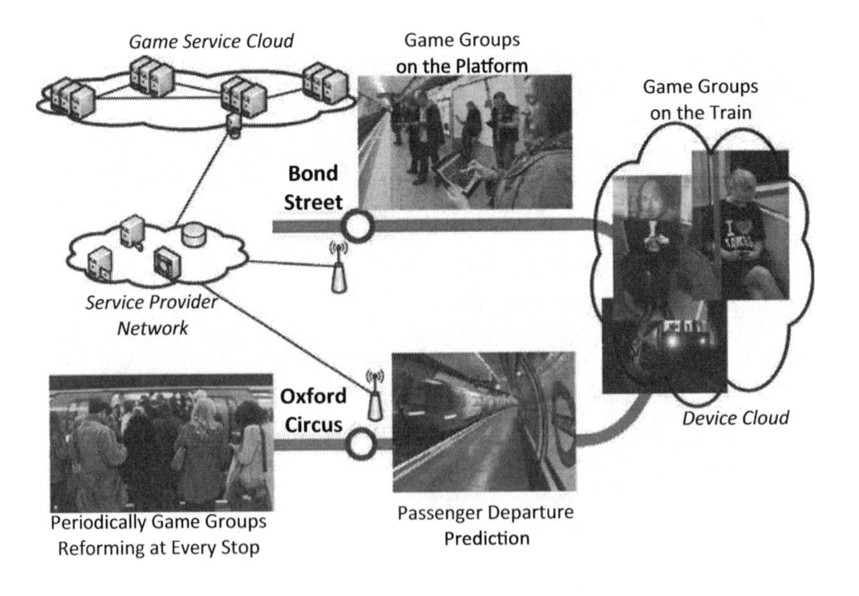

Figure 11.1: Overall scenario.

to, first-person-shooting (FPS) and racing. However, this comes with many challenges.

First, despite the fact that mobile gaming has been a dominant entertainment source in public transports, and the increasing number of mobile games available on major application markets, their adaptability as multi-player games is still in an early stage. As far as we are concerned, none of them is carefully designed for our targeted highly crowded public transport environments, as they are either single-player based or turn-wise multi-player, and these applications are usually built with suboptimal features, like for a particular platform and/or a network type with a mobile operator.

Second, in public transport, cellular network connectivity like 3G UMTS and even 4G LTE is not reliable at all. Figure 11.2 demonstrates the round trip time (RTT) between phones and the back-end game server via 3G/4G cellular connections. The experiments are undertaken on both the bus and metro scenarios in Seoul, South Korea, during non-peak hours. Since we are targeting to support multi-player games likes FPS and racing, the latency is required to be around 100-200ms and the jitter should be negligibly small. As shown in Figure 11.2, 3G network cannot serve the purpose for satisfactory gaming experience at all, and 4G network's unexpected high variance and high infrastructure costs may sometimes be a serious problem. Furthermore, we observe sudden link loss that may significantly impact the overall gaming experience. Thus, the challenge is to provide high-quality gaming experience to all passengers, without intermittent, abrupt response and interruptions. Our experiments confirm that a traditional client-server based gaming model via cellular networks, which may lead to an unreliable gaming connection and unsatisfactory user experience, is not feasible. If using broadcast, given that many gaming groups may potentially exist in a crowd, broadcasting from each end terminal would cause heavy co-channel inter-group interference.

Third, traditional cloud-based gaming services use game engines like [267], where mobile clients are only responsible for sending game-related commands to the cloud server. Although this centralized approach works on PCs, it suffers in our scenario where a stable Internet connectivity cannot be assumed, e.g., the train tunnel. Furthermore, the concept of realizing a gaming cloud Platform as a Service (PaaS) restricts the on-the-fly gaming behaviors of individual passengers, where the availability of the service is a prerequisite. Therefore, new technology that can cope with the *disrupted* mobile connections via Internet become strongly desired. Toward this end, the architecture of game development and communication methodology needs to be revised to consider the periodically disrupted manner of the Internet connection by the aid of a dynamic local cloud formed by smart devices.

Fourth, multi-player game play is further complicated by the mobility pattern of both the moving vehicles and passengers' different routines, e.g., the duration between two stops, the time window on the platform, where to board/get off the vehicle, etc.

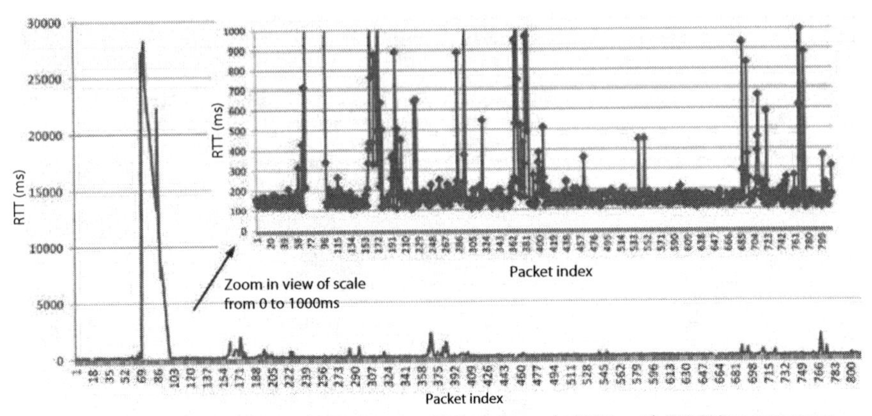

(a) Traveling on a bus with 3G network — RTT reaches up to 28.27s, with 738.518ms on average (standard deviation 3.147s) and 5% packet loss.

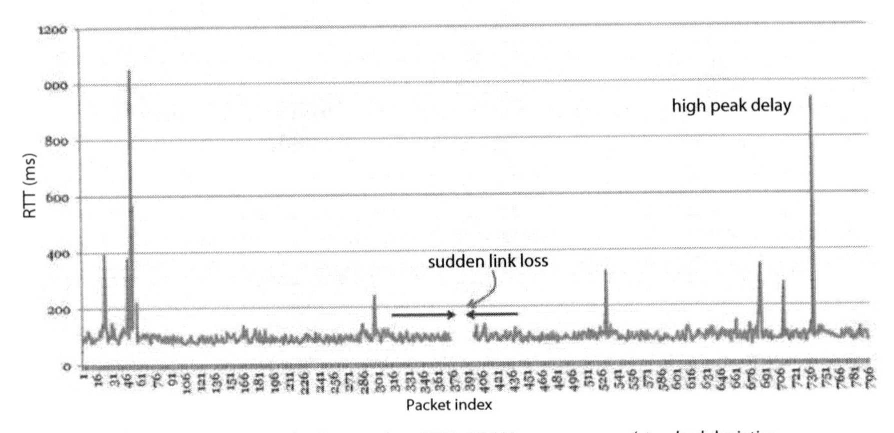

(b) Traveling in the metro with 4G network — RTT is 97.312ms on average (standard deviation 47.013ms) and 0.5% packet loss. Better performance but with sudden link loss and high peak delay.

Figure 11.2: Results of packets' RTT indicate 3G/4G cellular networks cannot serve the purpose for a satisfactory gaming experience because of high latency, unexpected high variance, and sudden link loss.

We developed a Bluetooth-based neighbor discovery application in an Android smartphone, where a tester carries the handset traveling in two fixed routes of Beijing Underground everyday from June to September 2012. As shown in Figure 11.3, experimental results confirm that the considered public transport scenarios indeed offer a unique opportunity for travelers to participate in game play according to their gaming preferences. It is very important to notice that the Bluetooth's sensing range is much shorter than WiFi's,

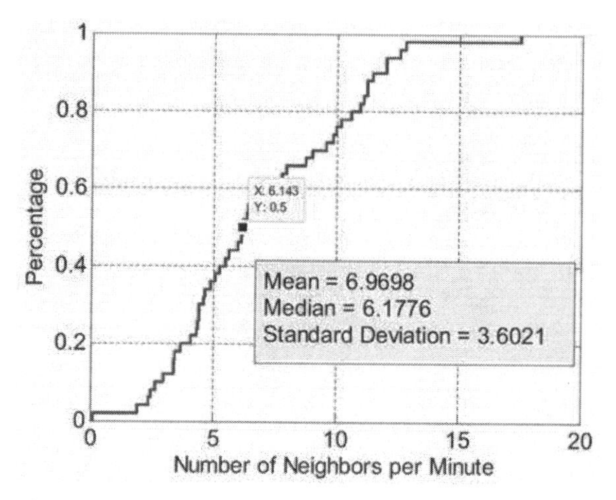

(a) The ECDF (empirical cumulative distribution function) of the number of newly discovered (Bluetooth enabled) neighbors per minute — On average nearly 7 new passengers are discovered per minute, which means that 14 new passengers are discovered on average at each stop of Beijing Underground given that it takes approximately 2 minutes for the trip between stops.

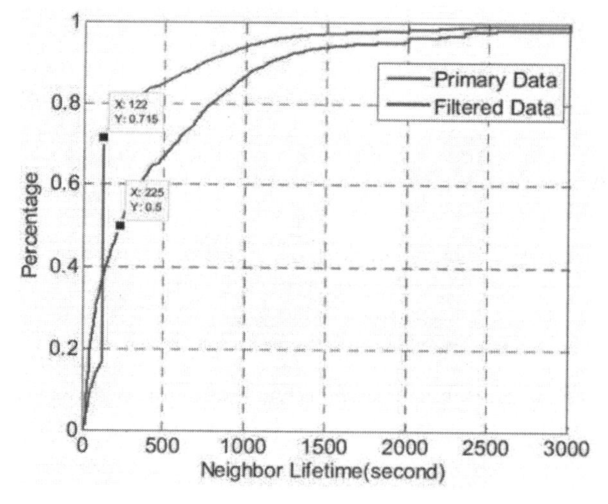

(b) The ECDF of the durations of neighbors staying within the communication range of Bluetooth — the average lifetime for an active passenger is 470.91s, i.e., on average a passenger will stay with the tester for 3.9 stops.

Figure 11.3: Studies of passenger dynamics in Beijing Underground suggest that there is ample opportunity to start/join a game play in the crowd. The experimental results also confirm that the considered public transport scenarios indeed offer a unique opportunity for travelers to participate in game play according to their gaming preferences.

therefore, we expect more passengers can be discovered via WiFi connections. However, the question still remains of how to optimally estimate the stability of an existing gaming group formed by passengers' devices, so that anyone's departure will not impact the overall gaming experience of others, and the newly arrived participants can accurately join the right game group and competitors among the crowd.

Finally, traditional ways like constant scanning and retrieving information from the cloud server in finding participants in the proximity may cause additional signaling overhead and energy efficiency issues. Furthermore, the conventional lengthy registration/sign-on/involvement process over the backend game server is obviously not applicable either. Therefore, a distributed game operation within the local crowd is strongly desired.

To address the above limitations and enable highly interactive gaming in crowded public transports, in this chapter we propose and prototype "CrowdMoG," the Crowd based Mobile Gaming platform. It provides users with cloud-based, context-aware automation for group forming and reforming when group members move. As a core functionality, CrowdMoG provides a location-aware *matchmaking process* to discover and match a passenger with his/her preferred game group. The backend cloud-based game servers offer players gaming group information according to the information retrieved from their mobile app as well as their social network profiles. Furthermore, the heavy lift computation like group forming and game owner (GO) selection are offloaded to the server in the cloud to guarantee minimum usage of the scarce battery and computation resources on the mobile client. Most importantly, since the numbers of platforms and trains can be large in a modern city, if the computation takes too long, the train will be gone before the players receive the relevant information. Therefore, the backend gaming servers have to be on the cloud which is resourceful in terms of computation power and knowledge.

The contribution of this chapter is threefold:

- We propose a completely new type of context-aware application scenario to enable highly interactive and opportunistic mobile gaming among nearby passengers in public transports, geared by the proposed CrowdMoG platform.

- We show the detailed component design of CrowdMoG to achieve the required functionalities, that include a few key design elements for gaming group discovery in aware of the passenger dynamics and related social information, and runtime game operation in aware of network latency issues: Cloud-Based Gaming Services, Cloud Manager, Peer Manager, Game Session Manager, Network Protocol Manager, Group Manager, and Game Feature Extractor.

- Finally, a prototype is developed and shown to demonstrate the gaming service among crowds.

The rest of the chapter is organized as follows. First we describe the potential scenario which frames key challenges in building such a system. Then, we elaborate the design of CrowdMoG. After the prototype is introduced, we conclude the chapter.

11.2 A CrowdMoG Use Case Scenario

In this section, we describe a motivating scenario for the considered opportunistic crowd-based gaming in public transports. Patrick from France boards a metro train departing for the Louvre in Paris. Being a regular commuter, he knows that the journey will take approximately half an hour, and thus decides to kill the time by playing a racing game on his Android tablet. With the in situ gaming functionality supported by CrowdMoG, he is able to race with nearby passengers in the same metro car. Patrick strongly favors this type of location-aware mobile gaming since he could play with people in the real world [263] (however, he may not know exactly who he is playing with), and more importantly, he would not experience unexpected performance degradation while playing due to the cellular network latency. When Patrick's game starts, the CrowdMoG platform on his tablet disseminates the gaming information, including the preferred game type, future period of stay on the route, the number of required players and their levels, etc., to all nearby passengers. Assume that this information is proactively received by two passengers in the same metro car who have installed the same game in their Google Nexus. Note that in practice, passengers frequently change their location (e.g., getting on and off) while commuting, and thus the maximum allowed time window to join a game is relatively very short (as analyzed in Section **Gaming Group Discovery** in Section 11.3.1). In such cases, just making an application layer protocol to probe for nearby peers is inefficient, since it will result in high battery drain while continuously searching for peers. Using our situation-aware matchmaking process as part of our CrowdMoG system, it helps identify only the GOs of the ongoing games if they are of similar interests as Patrick (rather than finding all game participants). Furthermore, CrowdMoG also performs seamless session handoff right after Patrick's departure, so that other game participants who were gaming with him are still able to maintain a satisfactory continuity, and one of them will become the new GO. Hence, the proposed system not only deals with intricacies related to the real-time, on-the-fly gaming, but also it solves the fundamental research problem of how to form a game group opportunistically at runtime among co-located travelers.

In the following discussions, we similarly call a passenger behaving like "Patrick" as the GO, those travelers who are physically around Patrick and within the communication range as the "crowd," and finally once they join Patrick in a game play, they are called the participants of a gaming group. A crowd may consist of multiple co-located simultaneous gaming groups.

11.3 CrowdMoG Design

Figure 11.4 shows the component design of CrowdMoG including: the cloud on the upper side which contains the cloud-based gaming services (CBGS) component and four layered design on the bottom that runs as a middleware service on each smart device. As an overall system, it receives supports from both the cloud-based game services that interact with social media (like online social networks) and provides the support for CrowdMoG and gaming applications on the smart devices. CrowdMoG consists of four layers built

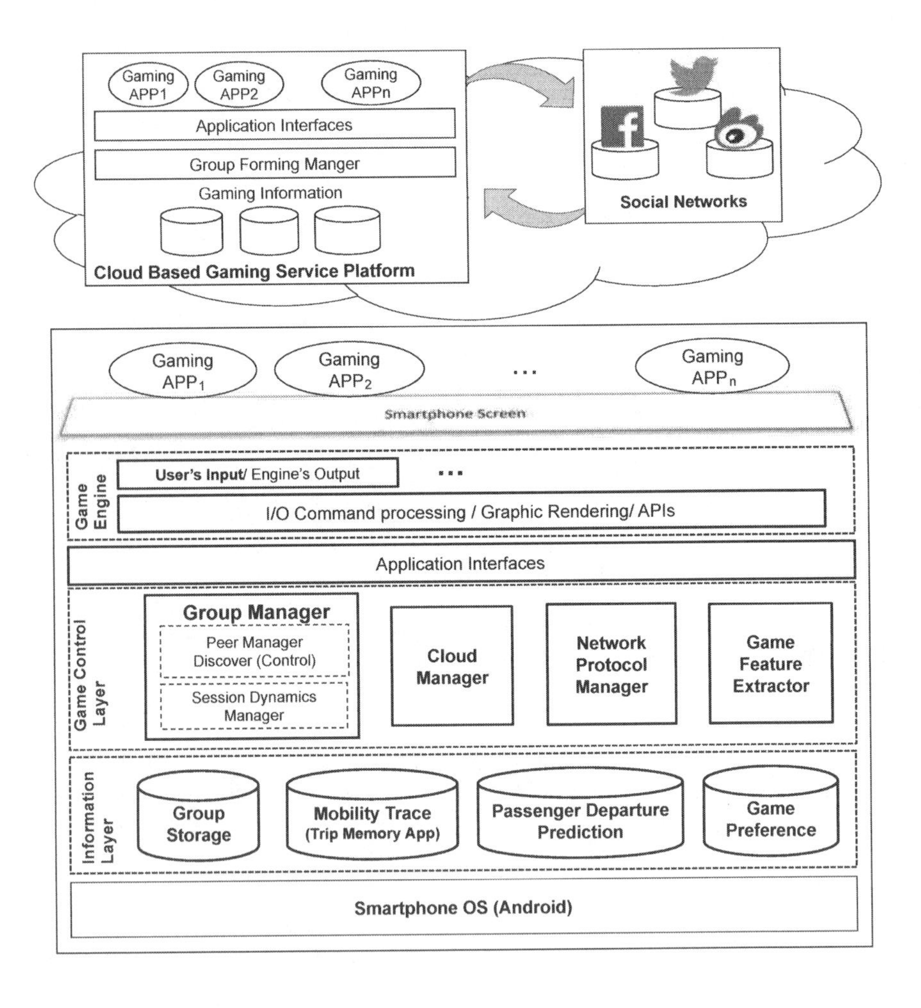

Figure 11.4: Architectural design of the CrowdMoG platform — the square on the bottom contains the part on the players' mobile devices while the cloud above contains cloud-based gaming services components.

on top of the smart device operating system (i.e., Android). The top layer leverages the existing open-source game engines (i.e., Andengine), responsible for I/O command processing, graphic rendering, and handling APIs. The next layer below consists of a handful of session APIs. These APIs are linked with the four modules in the Game Controller layer. It is primarily responsible for maintaining the gaming group session through the Group Manager as well as efficient communication with the Internet cloud (Cloud Manager) and local cloud (Network Protocol Manager). The session information is then passed to the Information layer, which maintains a set of repositories to store the gaming group information, passenger trajectories, etc. The fundamental functionalities associated with these key components are provided as follows.

11.3.1 Cloud-Based Game Services

The cloud-based game services (CBGS) maintain the information that is collected from each active player and gaming group which includes: the player's id, current location, past trajectory, existing gaming group, etc. Moreover, CBGS collaborate with existing social networking services like Facebook, Twitter, and SINA Weibo to retrieve other information including: players' profiles and their gaming preferences, etc. Based on information collected, CBGS determine how gaming groups can be established at a particular station, and whether the existing gaming groups in the subway car shall be reformed to accept newly joined passengers who just boarded, or to terminate the participation with those who just got off. Most importantly, CBGS nominate two GOs to all co-located players within a group. The reason is to minimize the chance of game disruption in case the only GO decides to get off the train but there is no other GO to serve the gaming group.

While considering to nominate GOs, CBGS predict all players' trajectories among which the owner of the longest one is nominated as the "executive" GO and the second longest as the "deputy" GO. The dead-reckoning concept [268] (extrapolation-based) is used here to estimate the new position of the player by predicting his/her trajectory through the most recent information update. Other possible techniques include the interpolation based techniques [269], however due to the imperfect prediction results and thus the errors deviating from the ground truth, it may significantly affect the game play experience at runtime. If the current executive GO is leaving the train, the deputy GO shall step in and therefore without expensive overhead on the communication and computation from the player's mobile device. After all, CBGS dynamically maintains two GOs for each group existing on the train and periodically updates two GOs and group members considering players on the train and platform.

Gaming Group Discovery: Passengers in public transport scenarios are behaving dynamically and their average period of stay indicates the possibility of playing multi-player games. Unlike in delay tolerant networks (DTNs) where each device aims to learn the mobility pattern of all nearby devices for information sharing [270, 271], it is not applicable for real-time gaming applications in our scenario due to the concerns of resource limitation on mobile devices. The challenging task of how to efficiently *detect* the nearby co-located devices, especially when considering their associated gaming preferences is aided by our proposed CBGS module which is much more resourceful in terms of computational power and information storage.

CBGS aid the learning of the mobility patterns of a crowd of passengers on the platform through offloading the computation from their own smart device. This is because given the fact that the period of a train stay on each platform is only on the scale of seconds (e.g., 20 seconds on average in Beijing Underground), it is almost impossible for a smart device to compute and compare the trajectory of all the passengers on the platform. However, it is worth mentioning that the larger the size of the crowd, the better diversity and opportunity for gaming passengers to join a preferable game play.

On the other hand, CBGS collect the gaming preferences of each player from their mobile device and social networks to aid the process of providing recommendations on games and groups. When certain passengers are predicted to join or leave a game, an optimal solution for group reforming is computed and provided by the CBGS to maintain the undisrupted gaming experiences for other players.

Furthermore, we limit the number of players in a gaming group to the maximum of 10 members in order to guarantee the essential information exchanges for the game in case the CBGS are not available to the crowd. The reason for conducting such a limitation is because of the fact that a unicast-based peer discovery by using Wi-Fi Direct shows that maximally 10 users can be detected in a 15-second time window by using a 1MHz channel of 900MHz band.

After all, the computation of group reforming and recommendations on games are all offloaded to the cloud, and all the computed results are prepared before the train arrives at the platform. Because the heavy lifting computation on group forming is offloaded to the cloud, the limited computation and battery resource on the player's mobile device can focus only on the game playing and necessary communication in an efficient manner.

11.3.2 Cloud Manager

Cloud manager on the player's mobile device takes responsibilities of initiating the connections between the CBGS and all players, while it also optimizes the time slot in which the client should be probing the CBGS with the optimized energy and communication efficiency. This is aided by the sensor-based *passenger departure prediction* (PDP) module and stored players' historical trajectory.

Passenger Departure Prediction: It predicts when a passenger is about to depart from the train and, therefore, leave the game. The prediction is supported by the hardware like motion sensors and accelerometers on mobile devices, and PDP automatically alerts the GO about this departure. Based on the received alerts from multiple passengers, the GO can further predict that the train might be arriving soon. Based on this prediction, the cloud manager is able to maximize the bandwidth efficiency by only probing the CBGS when the train is about to arrive at the platform where the cellular signal is available.

11.3.3 Group Manager

This module maintains the gaming group information like the type of game being played and number of participants as group members. During the time when the CBGS are not available, the GO's Group Manager module distributes this information to the new passengers to aid their joining decision, which is eventually performed by the Peer Manager. When the GO or group members leave the group without being successfully predicted, Session Dynamics Manager will provide essential support for maintaining the game group structure until the next available group reforming recommendation is received from CBGS.

The group information will be updated when existing passengers join/leave the game, and it will be further uploaded to the CBGS by the GO whenever the Internet connection is available. Note that the group information will be updated periodically at runtime on the mobile side, therefore we only store it in a memory-based device cache (without persistency guarantees). As discussed earlier, in order to provide satisfactory latency performance, the size of the group is restricted for the purpose of imposing a hard requirement on the maximum allowed number of participants admitted in the game.

11.3.3.1 Peer Manager

While recommending peer-of-interest is the primary functionality of the CrowdMoG platform on CBGS, the Peer Manager module performs the peer detection to associate the end users with all interested nearby participants via recommended gaming groups, without being impacted by any side effect of high discovering latency. As part of the situation-aware matchmaking process, peer detection is only performed by the GO, who initializes the game at the very beginning, or the deputy GO after the first GO's departure, rather than finding all devices by each passenger individually. As described above, CBGS provides each player with relevant gaming group member information and recommended GOs. This flexible functionality enables every player to associate himself/herself only with their preferred game play, given the fact that potentially many different types of games co-exist in a crowd. The role of peer manager is to follow the support from CBGS and connect with the interested gaming group (via its members or GOs).

Furthermore, interacting with the Session Dynamics Manager, Peer Manager enables the group association only if the participants are mutually trustworthy and their size does not exceed a threshold (see Section **Network**

Protocol Manager, Section 11.3.4). It provides an opportunity in case a player wants to interact with someone he/she has played earlier in a game and makes sure the admitted new player will not affect the existing game performance. The degree of trust can be quantized by parameters like average gaming session length between two players.

11.3.3.2 Session Dynamics Manager

Since our targeted application scenario is associated with a highly dynamic crowd, maintaining the dynamics of the game sessions becomes extremely important. This module serves as part of the situation-aware matchmaking process, to provide passengers a unique way to attain continuous session management.

In the case of the executive GO leaving the game unexpectedly, the deputy GO's session dynamics manager broadcasts to the group members about his/her new ownership of the game group; thereafter the deputy GO performs as the executive GO until a better recommendation on group reforming is received from the CBGS. On the other hand, if a group member leaves the game unexpectedly, it is comparatively easier to maintain the group structure.

Following Patrick's example earlier, since he is a regular commuter, the CrowdMoG platform running on his smart device periodically records his travel pattern using the trip memory application [262, 272], which has been prototyped as a proof-of-concept. It is an Android application that tracks the commuter's traveling path and logs the surrounding events extracted from the ambient sensors. The collected trip information (stored in the mobility trace repository of the Information layer) is then distributed to the GO, and the corresponding Session Dynamics Manager coordinates the session handover well before his/her departure through the Network Protocol Manager. In this way, it facilitates the mobility pattern overhearing inside the group, and ultimately helps all participants achieve the continuity of the game play.

11.3.4 Network Protocol Manager

Recently, overhearing the broadcast packets has been applied in a typically tethered connection type, to support the traditional server-client model in a local video streaming [273]. It aims to overcome the shortages of prediction based techniques, but comes with shortages at the expense of heavy energy drain on 3G/4G, heavy co-channel inter-group interferences [273], and without the support of session handover upon a participant's departure and new members joining in.

The main drawback of using 3G/4G-based cellular networks to facilitate crowd gaming in public transports is due to its poor wireless link quality as demonstrated in Figure 11.2. This problem can be more severe in some cases where infrastructure is not available (like in the tunnels of London Underground). Thus, a traditional client-server model is not feasible. Meanwhile, the proposed broadcasting based protocols like WiFi-Direct [274] suffer from injecting significant amounts of co-channel interference within a group, even under a reasonable size of 50 passengers (the usual case during peak hours). Toward this end, our Network Protocol Manager aims to reduce the number of packet exchanges in the crowd while improving the latency performance of the network in real time.

Instead of communicating with the cloud-based game server for each game packet (e.g., mobility and game information), we operate the CrowdMoG platform in a two-phase mechanism. Each participant unicasts the game packet to the GO in the first phase, and then the GO broadcasts the collected set of information to the crowd. In this way, the communication only happens within the local game group, and the GO is the only one to broadcast the packets, while *overhearing* by all participants. Furthermore, unlike traditional prediction-based methods (e.g., interpolation [269] and extrapolation [268]) that estimates other characters' movements by using their past information, our Network Protocol Manager accurately disseminates the gaming information among the game group. This mitigates the problem caused by poor cellular network connectivity, and helps distribute the *exact* gaming information among all participants, rather than using any prediction-based methods that aim to estimate their movements. This is feasible in that the size of the group remains relatively stable during the travel time between two stops.

Finally, because the size of the group highly affects the performance of network overhearing, this module aims to dynamically estimate an optimal *threshold*, defined as the maximum number of participants in a group, beyond which the interference may deteriorate the protocol performance. This threshold is quantized by parameters like the received signal-to-noise-plus-interference-ratio (SINR) at runtime.

11.3.5 Game Feature Extractor

This module is primarily responsible for collecting and analyzing the smart device user's gaming preferences data that is produced on the mobile device, including the preferred game type and role, the average duration of the game play, the preferred type of competitors, etc. These pieces of information are all periodically collected and summarized by the Game Feature Extractor when the Internet connection is not available and then, this module is expected to perform runtime analysis to extract the gaming preferences of the user, and later stored in the Game Preference repository in the Information layer.

11.4 Prototype — *Phage*

"*Phage*" is developed as a proof-of-concept to demonstrate the advantage of the proposed gaming platform. It is a PvP (player vs. player) real-time strategy game that is very easy and enjoyable to play. The game is about

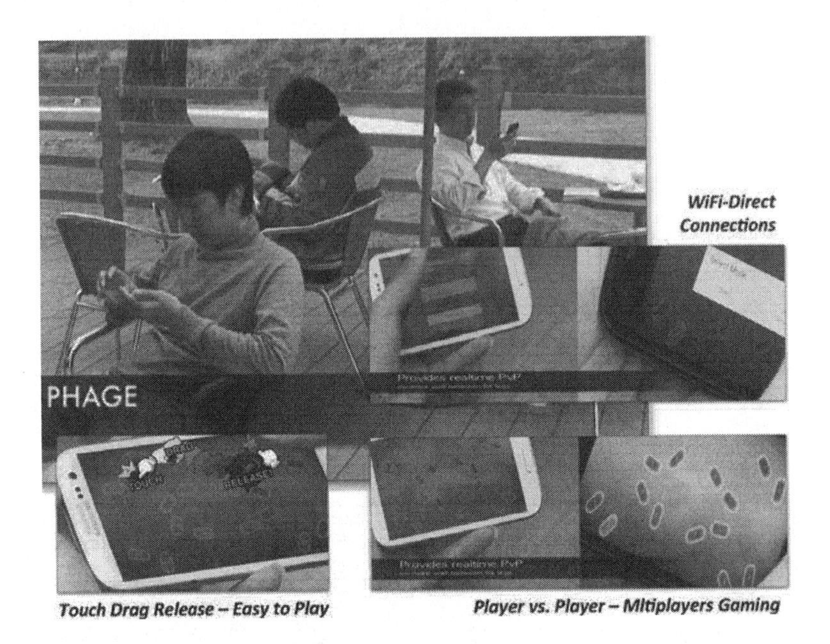

Figure 11.5: Screenshot of the "Phage."

multiple species of viruses aiming to infect as many cells as possible. In order to infect a cell, players first touch and drag over cells that are already occupied by his/her "virus" in an attempt to assemble a relatively larger quantity of "virus" army, and then release the finger at another cell to direct the army to defeat those occupying "viruses" from another species. Eventually, the winning species should infect all cells by defeating every other species of "virus."

The game is easy to play because it uses touch interface only on which a single movement is conducted by a simple touch-drag-release sequence. Players can operate on the touchscreen to: (1) touch the factions to choose the cells in which the occupying viruses are selected, (2) drag and hover over other cells in your faction to fortify the cell or aim at others to attack, (3) and then release to make the fortification or attack happen. In the team-wise multi-player case, two players can also communicate with each other to collaborate on a joint operation to defeat a third player.

"*Phage*" supports the 3G communication with CBGS and the WiFi-Direct for ad hoc multi-player gaming. As demonstrated in Figure 11.5, "*Phage*" is very suitable for a group of temporarily co-located passengers to enjoy some moments of fun during their time of travel. On the one hand, the feature that has WiFi-Direct to connect with other players provides the game with advantages in the underground tunnel. Because even when a public telecommunication network is not available, players can still enjoy the current game session via WiFi-Direct connection. On the other hand, it also supports the 3G communication with the CBGS while the train is on the platform where Internet infrastructure can be assumed. This provides the game application with the ability to connect with the SINA Weibo, the most visited microblogging website in China, where the players publish their scores and the CBGS retrieves the gaming preference information.

At the current stage of development, although only one game has been implemented based on proposed platform, it demonstrates the applicability of the overall platform in providing cloud-based context-aware crowd gaming service.

11.5 Summary

Multi-player gaming on smart devices is increasingly catching attention, particularly in public transport environments where people are looking forward to playing rich contextual games while traveling. In this chapter, we proposed CrowdMoG, a crowd-based mobile gaming platform that utilizes the cloud resource as well as crowd resource to enable highly interactive and opportunistic mobile gaming in such scenarios. The goal is to leverage the unlimited computation power of the cloud, information on the social network, and the mobile devices in the vicinity to achieve a distributed on-line gaming experience in real life.

Specifically, in this chapter we proposed a design of the CrowdMoG platform with key design elements including: Cloud-Based Gaming Services, Group Manager, Cloud Manager, Network Protocol Manager, and Game Feature Extractor. Supported by many modules, the group forming process offers a unique functionality to discover and match the nearby passengers on the move according to their associated gaming preferences, and provides the smooth session handover to enable the continuity of existing game plays when participants leave the game due to different mobility patterns.

The *Phage*, a functioning vertical prototype, has been developed to demonstrate the advantage of the proposed gaming platform and cloud-based crowd architecture. In addition to the multi-players' gaming experience, the Phage provides passengers with convenient ad hoc communications via WiFi-Direct connection. The game preference extractor, which is the fundamental support for game group forming, is also implemented to analyze the favorite based on a player's profile on social network.

Our investigation aims to demonstrate the applicability of context-aware crowd gaming in underground transportation where the public telecommunication network is often not reliable. The proposed architecture enables the gaming app to behave differently according to the availability of the cellular network. On one hand, the application can automatically form/join a game group based on the information retrieved when the cellular connection is available. On the other hand, when the cellular network is not available, a local device cloud is sufficient to maintain the gaming communication.

Chapter 12

Mobile Cloud Computing Enabled Emerging CPS Applications

Chi Harold Liu

Beijing Institute of Technology, China

CONTENTS

The previous chapters present two important application areas of CPS. In fact, there are many emerging CPS applications to be considered on such platforms, e.g., *crowd computing*, *mobile social networking applications*, *location-based services*, *sensory-based applications*, etc. Realization of these applications brings many constraints for researchers and developers like task offloading to reduce energy consumption on a device and execution time, communication bandwidth, security and privacy, etc. In this chapter, we name some representative MCC-enabled CPS applications in the following four areas: *education*, *office automation*, *healthcare*, and *mission-critical applications.*

12.1 Education

Since CPS is based on the connection of common "objects" to the Internet, inherently it provides powerful possibilities to enhance the traditional learning experience, and brings a high degree of tangibility to the learning process, especially with user-centered characteristics [275]. Specifically, powered by MCC technologies, mobile learning represents an expansion of the learning environment. Compared to conventional e-Learning, MCC-enabled CPS architecture and associated key technologies enable mobile learning to provide full accessibility to lecture contents and acquire knowledge without the constraint of specific place, time, or devices. Furthermore, the personal smart devices can establish better interactions among students, and between students and lecturer.

A most recent implementation of an intelligent mobile cloud education system is reported in [276]. The system provides four main functionalities of Learning-Assessment-Communication-Analysis (LACA) based on the IaaS architecture to enable learners to access the learning cloud anytime anywhere on various mobile devices. Its primary motivation is that the role of teaching and learning has changed from a teacher-centered approach to a community-centered approach, and from information transfer to knowledge creation. Considering the type of MCC-oriented CPS architecture, `Internet Cloud` can be most suitable for educational, mobile applications. Key challenges in this area are the limited device battery due to the power-consuming nature of video streaming applications, limited communication bandwidth, security, and privacy due to online mobile exams, etc.

12.2 Office Automation

Smart environments, such as smart home, smart office, and smart plant, are envisioned as one of the most profound impacts within the concept of CPS, making full human interactions with surrounding systems a pleasant experience. Specially, in the field of office automation, the rapid growth of the usage of smart devices and the rise of cloud computing have motivated practitioners to provide solutions that can benefit the user not only on the level of everyday office activities such as "driverless printing" [277], but also on the enterprise level [278]. For the latter, Cisco developed and deployed a cloud-based selling management system [278] to its worldwide field users, to not only provide uniform representation transformation to ensure that the same enterprise content is natively accessible from mobile devices, but also take

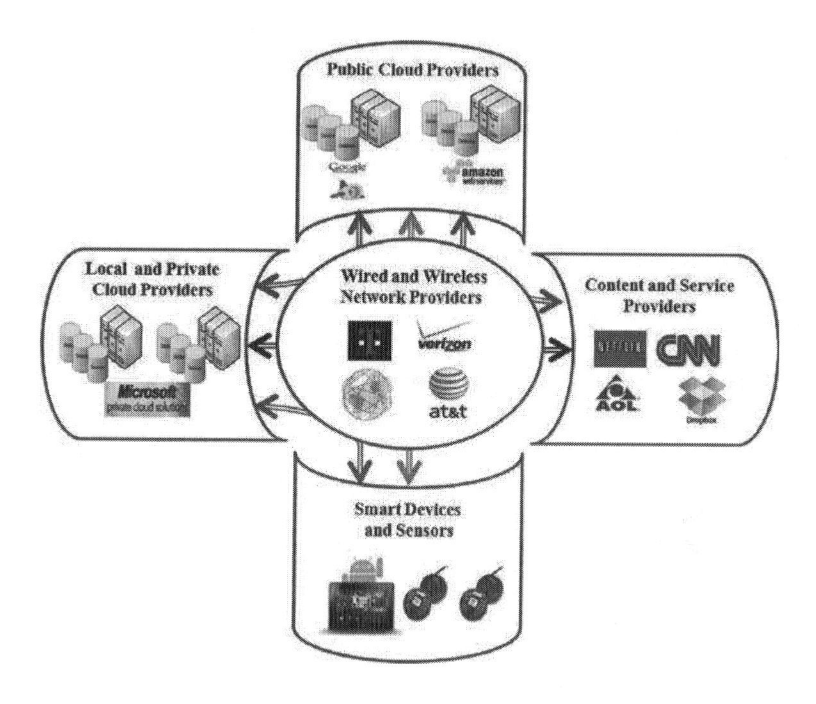

Figure 12.1: MCC ecosystem and its important stakeholders.

over the heavy lifting such as: access authentication, tracking, and business analytics. Uniform user experience across Apple iOS, Blackberry, and Android devices is provided through a thin smartphone native client. Therefore, "anywhere & anytime" information retrieval is ensured. More importantly, global collaborative communities are formed among colleagues because it allows field personnel with the same interests to share updates on strategies or exchange important information in order to make the next sale cycle easier.

In the scenarios of office automation application within a specific geographical range, like printer, projector, etc., `Device Cloud` provides more efficient connectivities. In addition, `Internet Cloud` remains the most suitable solution for field agents. The main challenges faced here are: (a) the authentication on access from mobile devices takes significant effort, and (b) heterogeneity of mobile devices with different user interfaces causes problems when the content and user interface are updated.

12.3 Healthcare

One of the main and emerging CPS applications is m-health [279]. Realization of m-health applications will bring many challenges like real-time processing and storing huge amounts of patient data, privacy protection and security of medical data, power-awareness of sensors, bandwidth constraints, etc.

Through smart devices in a CPS sensory environment, MCC is able to provide patients with easier accessibility to critical healthcare. ROCHAS [280] assists the healthcare service between patients and doctors by adopting wired or wireless communications, and a cloud-assisted healthcare infrastructure. In addition, a low-cost robot (as the end device) is used to provide mobile multimedia communications and intelligent speech recognition, so as to help the patient to connect and communicate with other family members as well.

In the scenarios of healthcare monitoring, `Hybrid Cloud` is the solution. The `Device Cloud` provides "anywhere" connectivity to enable real-time monitoring and managing a range of medical conditions. The `Internet Cloud` provides "anytime" resourceful support to generate accurate estimation on a patient's status. On the other hand, the `Internet Cloud` is a more resourceful pool to find aids for remote diagnostics, and `Device Cloud` is the only efficient way to connect and communicate in the case of large-scale disaster management.

12.4 Mission-Critical Applications

In a critical mission-based field operation, like disaster management, emergency rescue mission, etc., maintaining local communications and information sharing in real-time are very critical for the crowds (i.e., soldiers) in the front line. The `Device Cloud` architecture provides a way that not only allows the crowds in the front line to communicate and share information, but also to optimize the computational and battery resources among them for the purpose of maximizing the lifetime of the teams as a whole.

Recently, CPS has obtained rapid development and has a significant impact on the military field. The Defense Advanced Research Projects Agency (DARPA) recently conducted a preliminary field test on their software program called Content Based Mobile Edge Networking (CBMEN) [281] which allows soldiers in the battlefield to share intelligence without syncing to a server by using the equipped smart devices.

From the technical point of view, DARPA's effort is to leverage the available connections like cellular, WiFi, or Rifleman Radios (as the access network in a CPS architecture) to create secure front line cloud storage services that provide battlefield content with decreased latency and increased availability.

On the one hand, the updates are automatically replicated and shared as long as the troops are within the connection range, and the tactical cloud grows and diminishes as soldiers move in and out of range of each other. On the other hand, when the troops come back to the base where connections becomes available, all the intelligence can be synced to the central server and processed on the infrastructure-based cloud.

The challenges faced here are: (a) limited battery of the mobile device, (b) inefficient information exchange, (c) limited bandwidth of the connection, and (d) heterogeneity of the mobile devices and network connection.

12.5 Summary

These MCC-enabled CPS applications are not just simply the mobile version of PC applications. Their mobility and dynamic nature bring something really challenging and important for researchers and practitioners to tackle. Traditionally, all computation was run on fixed PCs where energy consumption and computing performance of a PC's application were not constrained. However, this is completely different when it comes to the era of MCC with limited mobile battery and CPU power. In addition, the heterogeneity in an MCC environment, OS, wireless connections, etc., requires the MCC application to be much more adaptive and dynamic. Also, in contrast to a fixed PC, a mobile phone is a very personal belonging, therefore extra security and privacy protection are highly expected.

Chapter 13

Conclusion

Chi Harold Liu and Jianxin Zhao
Beijing Institute of Technology, China

Yan Zhang
Simula Research Laboratory and University of Oslo, Norway

Cyber Physical Systems (CPS) represent an evolution in computerized interconnectivity. It integrates the dynamics of the physical processes with those of the software and communication, providing abstractions and modeling, design, and analysis techniques for the integrated whole. It has become the new research focus for worldwide industry, academia, and government agencies. CPS faces many challenges, including CPS-specific ones such as privacy, participatory sensing, data analytics, GIS-based visualization, cloud computing, and those standard WSN challenges, e.g., architecture, energy efficiency, security, protocols, and quality of service. To deal with these challenges, in this book, we analyze CPS in detail from different aspects, including its architectures, protocols, and applications.

In Chapter 2, we discussed the overall CPS architecture by analyzing different CPS applications from the bottom up and extracting the common characters that form a vertical structure. The next four chapters presented different important aspects of this structure. Chapter 3 presented mobile sensing platforms and their applications in interrelated paradigms, and highlighted and briefly discussed different types of mobile sensing platforms and the functionalities they can offer. In Chapter 4, we discussed the naming, addressing, and profile services of CPS and proposed a middleware component to meet the requirements of dynamic applications, and sensors/actuators deployment/configurations across different platforms. Chapter 5 presented a context-aware sensor search, selection, and ranking model, which addresses the challenge of efficiently selecting a subset of relevant sensors out of a large set of sensors with similar functionality and capabilities. In Chapter 6, we considered various topics in the energy management of CPS and proposed a novel energy-efficient framework.

The next part presented some vital technologies that make CPS possible. Chapter 7 presented the fundamental networking technologies of CPS. Chapter 8 focused on the machine-to-machine communications for CPS, specifically on open technologies such as IPv6-based solutions that can be integrated into IoT and enable wireless sensor communications. Chapter 9 discussed the state-of-the-art MCC research activities that aim to enhance the capabilities of resource-constrained smart devices in CPS sensory environments. With its immense and complex information perceiving and processing capabilities, MCC technologies are emerging to empower diverse CPS applications by providing ubiquitous and rich functionalities.

Finally, Chapters 10, 11, and 12 presented a few representative CPS applications, including connected healthcare, gaming in public transport crowds, and a series of MCC-enabled emerging CPS applications. Built on previous architecture and technologies, these application fields fully demonstrate the great potential of applying CPS in public life.

Although there are still many research challenges in CPS, we believe that CPS will finally be extended to individual and home users, facilitating life for everyone with ubiquitous networks and strong connections among all things. CPS will also extend communications beyond the traditional boundaries. As more devices are added to communication networks, ubiquitous networks will take shape and grow. The gradual establishment and improvement of technology will inevitably enlarge the scope of CPS applications and lower deployment costs so that they will finally become part of daily life.

References

[1] W. Wolf. Cyber-physical systems. *Computer*, 42(3):88–89, March 2009.

[2] Kyoung-Dae Kim and P. R. Kumar. Cyber-physical systems: A perspective at the centennial. *Proceedings of the IEEE*, 100 (Special Centennial Issue):1287–1308, 2012.

[3] R. R. Rajkumar, I. Lee, L. Sha, and J. Stankovic. Cyber-physical systems: The next computing revolution. In *Proceedings of the 47th Design Automation Conference*, pages 731–736. ACM, 2010.

[4] L. Atzori, A. Iera, and G. Morabito. The Internet of Things: A survey. *Comput. Netw.*, 54:2787–2805, Oct. 2010.

[5] J. Zheng, D. Simplot-Ryl, C. Bisdikian, and H. T. Mouftah. The Internet of Things. *Special Topic in IEEE Comm. Mag.*, Nov. 2011.

[6] Daniele Miorandi, Sabrina Sicari, Francesco De Pellegrini, and Imrich Chlamtac. Internet of things: Vision, applications and research challenges. *Ad Hoc Networks*, 10(7):1497–1516, 2012.

[7] Jianhua Shi, Jiafu Wan, Hehua Yan, and Hui Suo. A survey of cyber-physical systems. In *IEEE WCSP'11*, pages 1–6, 2011.

[8] Jiann-Liang Chen, Ming-Chiao Chen, Chien-Wu Chen, and Yao-Chung Chang. Architecture design and performance evaluation of RFID object tracking systems. *Computer Communications*, 30(9):2070–2086, 2007.

[9] Chris M Roberts. Radio frequency identification (rfid). *Computers & Security*, 25(1):18–26, 2006.

[10] Kevin Ashton. That 'internet of things' thing. *RFID Journal*, July 2009.

[11] Jean-Philippe Vasseur and Adam Dunkels. *Interconnecting smart objects with IP: The next internet.* Morgan Kaufmann, 2010.

[12] INFSO d.4 networked enterprise & RFID INFSO g.2 micro & nanosystems. In *Co-operation with the working group RFID of the ETP EPOSS*, 2008.

[13] P. Kinney. ZigBee technology: Wireless control that simply works. In *Communications design conference*, volume 2, 2003.

[14] B. Mukherjee, J. Yick, and D. Ghosal. Wireless sensor network survey. *Computer Networks*, 52(12):2292–2330, 2008.

[15] I. Toufik, S. Sesia, and M. Baker. LTE: the UMTS long term evolution. *Wiley Online Library*, 2009.

[16] Haiyan Luo, Song Ci, Dalei Wu, Nicholas Stergiou, and Ka-Chun Siu. A remote markerless human gait tracking for e-healthcare based on content-aware wireless multimedia communications. *IEEE Wireless Communications*, 17(1):44–50, 2010.

[17] N. Bui and M. Zorzi. Health care applications: A solution based on the internet of things. In *The 4th International Symposium on Applied Sciences in Biomedical and Communication Technologies*, 2011.

[18] J. Paradells and C. Gomez. Wireless home automation networks: A survey of architectures and technologies. *IEEE Comm. Mag.*, 48(6):92–101, 2010.

[19] P. Corke, T. Wark, R. Jurdak, W. Hu, P. Valencia, and D. Moore. Environmental wireless sensor networks. *Proceedings of the IEEE*, 98(11):1903–1917, 2010.

[20] F.-Y. Wang. Parallel control and management for intelligent transportation systems: Concepts, architectures, and applications. *IEEE Transactions on Intelligent Transportation Systems*, 11(3):630–638, 2010.

[21] B. McMillin, C. Gill, M.L. Crow, F. Liu, D. Niehaus, A. Potthast, and D. Tauritz. Cyber-physical systems distributed control: The advanced electric power grid. *Proc. of Electrical Energy Storage Applications and Technologies*, 2007.

[22] D. R. Garrison. *E-learning in the 21st century: A framework for research and practice.* Taylor & Francis, 2009.

[23] Jayavardhana Gubbi, Rajkumar Buyya, Slaven Marusic, and Marimuthu Palaniswami. Internet of things (iot): A vision, architectural elements, and future directions. *Future Generation Computer Systems*, 29(7):1645–1660, 2013.

[24] Andrew T. Campbell, Shane B. Eisenman, Nicholas D. Lane, Emiliano Miluzzo, Ronald Peterson, Hong Lu, Xiao Zheng, Mirco Musolesi, Kristof Fodor, and Gahng-Seop Ahn. The rise of people-centric sensing. In *IEEE Internet Comp. Special Issue on Sensor Networks*, 2008.

[25] N.D. Lane, E. Miluzzo, Hong Lu, D. Peebles, T. Choudhury, and A.T. Campbell. A survey of mobile phone sensing. *Communications Magazine, IEEE*, 48(9):140–150, Sept. 2010.

[26] Paul Lilly. Mobile devices to outnumber global population by 2017. http://hothardware.com/News/Mobile-Devices-To-Outnumber-Global-Population-By-2017/ [Accessed on: 2013-08-06].

[27] Arkady Zaslavsky, Charith Perera, and Dimitrios Georgakopoulos. Sensing as a service and big data. In *International Conference on Advances in Cloud Computing (ACC-2012)*, pages 21–29, Bangalore, India, July 2012.

[28] Harald Sundmaeker, Patrick Guillemin, Peter Friess, and Sylvie Woelfflé. Vision and challenges for realising the internet of things. Technical report, European Commission Information Society and Media, March 2010. `http://www.internet-of-things-research.eu/pdf/IoT_Clusterbook_March_2010.pdf` [Accessed on: 2011-10-10].

[29] Cosm. Cosm platform, 2007. `https://cosm.com/` [Accessed on: 2012-08-05].

[30] CoAP, http://tools.ietf.org/html/draft-ietf-core-coap-04.

[31] Standardized machine-to-machine (M2M) software development platform, white paper, InterDigital, Inc.

[32] Machine-to-machine (M2M) — the rise of the machines, white paper, Juniper Networks, Inc.

[33] Charith Perera, Arkady Zaslavsky, Peter Christen, and Dimitrios Georgakopoulos. Sensing as a service model for smart cities supported by internet of things. *Transactions on Emerging Telecommunications Technologies (ETT)*, 2014.

[34] Xiang Sheng, Jian Tang, Xuejie Xiao, and Guoliang Xue. Sensing as a service: Challenges, solutions and future directions. *Sensors Journal, IEEE*, 13(10):3733–3741, 2013.

[35] Prem Prakash Jayaraman, Charith Perera, Dimitrios Georgakopoulos, and Arkady Zaslavsky. Efficient opportunistic sensing using mobile collaborative platform. In *9th IEEE International Conference on Collaborative Computing: Networking, Applications and Worksharing (COLLABORATECOM)*, Austin, Texas, United States, October 2013.

[36] Edward A. Lee. Cyber-physical systems: A rehash or a new intellectual challenge?, June 2013. Invited Talk in the Distinguished Speaker Series, sponsored by the IEEE Council on Electronic Design Automation (CEDA) held at the Design Automation Conference (DAC), Austin, Texas.

[37] T. Kakantousis and V. Kalogeraki. A mobile platform for managing mobile MapReduce participatory sensing data. In *Applications and the Internet (SAINT), 2012 IEEE/IPSJ 12th International Symposium on*, pages 196–201, 2012.

[38] T. Choudhury, S. Consolvo, B. Harrison, J. Hightower, A. LaMarca, L. Legrand, A. Rahimi, A. Rea, G. Bordello, B. Hemingway, P. Klasnja, K. Koscher, J.A. Landay, J. Lester, D. Wyatt, and D. Haehnel. The mobile sensing platform: An embedded activity recognition system. *Pervasive Computing, IEEE*, 7(2):32–41, 2008.

[39] Woosuk Lee, Bodhi Priyantha, Ted Hart, Gerald DeJean, Yan Xu, and Jie Liu. The CLEO mobile sensing platform. In *Proceedings of the 10th ACM Conference on Embedded Network Sensor Systems*, SenSys '12, pages 371–372, New York, NY, USA, 2012. ACM.

[40] Teemu Laukkarinen, Jukka Suhonen, and Marko Hnnikinen. An embedded cloud design for internet-of-things. *International Journal of Distributed Sensor Networks*, 2013:13, 2013.

[41] A. Sashima, Y. Inoue, T. Ikeda, T. Yamashita, and K. Kurumatani. Consorts-s: A mobile sensing platform for context-aware services. In *Intelligent Sensors, Sensor Networks and Information Processing, 2008. ISSNIP 2008. International Conference on*, pages 417–422, 2008.

[42] Arkady Zaslavsky, Prem Prakash Jayaraman, and Shonali Krishnaswamy. ShareLikesCrowd: Mobile analytics for participatory sensing and crowd-sourcing applications. *2013 IEEE 29th International Conference on Data Engineering Workshops (ICDEW)*, 0:128–135, 2013.

[43] Niels Brouwers and Koen Langendoen. Pogo, a middleware for mobile phone sensing. In *Proceedings of the 13th International Middleware Conference*, Middleware '12, pages 21–40, New York, NY, USA,, 2012. Proceedings of the 13th International Middleware Conference, Springer-Verlag New York, Inc.

[44] R.K. Ganti, Fan Ye, and Hui Lei. Mobile crowdsensing: Current state and future challenges. *Communications Magazine, IEEE*, 49(11):32–39, November 2011.

[45] B. Predic, Zhixian Yan, J. Eberle, D. Stojanovic, and K. Aberer. ExposureSense: Integrating daily activities with air quality using mobile participatory sensing. In *Pervasive Computing and Communications Workshops (PERCOM Workshops), 2013 IEEE International Conference on*, pages 303–305, 2013.

[46] H.A. Franke, F.L. Koch, C.O. Rolim, C.B. Westphall, and D.O. Balen. Grid-m: Middleware to integrate mobile devices, sensors and grid com-

puting. In *Wireless and Mobile Communications, 2007. ICWMC '07. Third International Conference on*, pages 19–19, 2007.

[47] Li Zhang, Qiang Wang, and Xijuan Shu. A mobile-agent-based middleware for wireless sensor networks data fusion. In *Instrumentation and Measurement Technology Conference, 2009. I2MTC '09. IEEE*, pages 378–383, 2009.

[48] Yong Sun and K. Nakata. An agent-based architecture for participatory sensing platform. In *Universal Communication Symposium (IUCS), 2010 4th International*, pages 392–400, 2010.

[49] M. Budde, M. Berning, M. Busse, T. Miyaki, and M. Beigl. The teco envboard: A mobile sensor platform for accurate urban sensing and more. In *Networked Sensing Systems (INSS), 2012 Ninth International Conference on*, pages 1–2, 2012.

[50] Dirk Trossen and D. Pavel. Nors: An open source platform to facilitate participatory sensing with mobile phones. In *Mobile and Ubiquitous Systems: Networking Services, 2007. MobiQuitous 2007. Fourth Annual International Conference on*, pages 1–8, 2007.

[51] V. Agarwal, N. Banerjee, D. Chakraborty, and S. Mittal. Usense – a smartphone middleware for community sensing. In *Mobile Data Management (MDM), 2013 IEEE 14th International Conference on*, volume 1, pages 56–65, 2013.

[52] A.S. Shirazi, C. Winkler, and A. Schmidt. Sense-sation: An extensible platform for integration of phones into the web. In *Internet of Things (IOT), 2010*, pages 1–8, 2010.

[53] M. Blackstock, N. Kaviani, R. Lea, and A. Friday. Magic broker 2: An open and extensible platform for the internet of things. In *Internet of Things (IOT), 2010*, pages 1–8, 2010.

[54] A. Purohit, Zheng Sun, F. Mokaya, and Pei Zhang. Sensorfly: Controlled-mobile sensing platform for indoor emergency response applications. In *Information Processing in Sensor Networks (IPSN), 2011 10th International Conference on*, pages 223–234, 2011.

[55] Ted Tsung-Te Lai, Wei-Ju Chen, Kuei-Han Li, Polly Huang, and Hao-Hua Chu. Triopusnet: Automating wireless sensor network deployment and replacement in pipeline monitoring. In *Proceedings of the 11th International Conference on Information Processing in Sensor Networks*, IPSN '12, pages 61–72, New York, NY, USA, 2012. ACM.

[56] German Castignani, Alejandro Lampropulos, Alberto Blanc, and Nicolas Montavont. Wi2me: A mobile sensing platform for wireless heterogeneous networks. In *Proceedings of the 2012 32Nd International Conference on Distributed Computing Systems Workshops*, ICDCSW '12, pages 108–113, Washington, DC, USA, 2012. IEEE Computer Society.

[57] D. Kharrat and S.S. Quadri. Self-registering plug-ins: an architecture for extensible software. In *Electrical and Computer Engineering, 2005. Canadian Conference on*, pages 1324–1327, 2005.

[58] Karl Aberer, Manfred Hauswirth, and Ali Salehi. Infrastructure for data processing in large-scale interconnected sensor networks. In *International Conference on Mobile Data Management*, pages 198–205. International Conference on Mobile Data Management, May 2007.

[59] Michael Compton, Corey Henson, Holger Neuhaus, Laurent Lefort, and Amit Sheth. A survey of the semantic specification of sensors. In *2nd International Workshop on Semantic Sensor Networks, at 8th International Semantic Web Conference,* October 2009.

[60] Michael Compton. Holger Neuhaus. The semantic sensor network ontology: A generic language to describe sensor assets. In *AGILE 2009 Pre-Conference Workshop Challenges in Geospatial Data Harmonisation*, 2009.

[61] Libelium Comunicaciones Distribuidas. libelium, 2006. `http://www.libelium.com/` [Accessed on: 2012-011-28].

[62] Charith Perera, Arkady Zaslavsky, Peter Christen, Ali Salehi, and Dimitrios Georgakopoulos. Capturing sensor data from mobile phones using global sensor network middleware. In *IEEE 23rd International Symposium on Personal Indoor and Mobile Radio Communications (PIMRC)*, pages 24–29, Sydney, Australia, September 2012.

[63] GSN Team. Global sensor networks project, 2011. `http://sourceforge.net/apps/trac/gsn/` [Accessed on: 2011-12-16].

[64] Charith Perera, Prem Jayaraman, Arkady Zaslavsky, Peter Christen, and Dimitrios Georgakopoulos. *Big Data and Internet of Things: A Roadmap for Smart Environments*, chapter Context-aware Dynamic Discovery and Configuration of 'Things' in Smart Environments, pages 215–141, Springer Berlin Heidelberg, 2014.

[65] Charith Perera, Prem Jayaraman, Arkady Zaslavsky, Peter Christen, and Dimitrios Georgakopoulos. Dynamic configuration of sensors using mobile sensor hub in internet of things paradigm. In *IEEE 8th International Conference on Intelligent Sensors, Sensor Networks, and Information Processing (ISSNIP)*, pages 473–478, Melbourne, Australia, April 2013.

[66] S. Hadim and N. Mohamed. Middleware: Middleware challenges and approaches for wireless sensor networks. *Distributed Systems Online, IEEE*, 7(3):1, March 2006.

[67] Charith Perera, Prem Prakash Jayaraman, Arkady Zaslavsky, Peter Christen, and Dimitrios Georgakopoulos. Mosden: An internet of things middleware for resource constrained mobile devices. In *47th Hawaii International Conference on System Sciences (HICSS)*, page n/a, Kona, Hawaii, USA, January 2014.

[68] Jin-Shyan Lee, Yu-Wei Su, and Chung-Chou Shen. A comparative study of wireless protocols: Bluetooth, UWB, ZigBee, and wi-fi. In *Industrial Electronics Society, 2007. IECON 2007. 33rd Annual Conference of the IEEE*, pages 46–51, 2007.

[69] D. Carlson and A. Schrader. Dynamix: An open plug-and-play context framework for android. In *Internet of Things (IOT), 2012 3rd International Conference on the*, pages 151–158, 2012.

[70] Kiev Gama, Lionel Touseau, and Didier Donsez. Combining heterogeneous service technologies for building an internet of things middleware. *Computer Communications*, 35(4):405–417, 2012.

[71] Luis Roalter, Matthias Kranz, and Andreas Moller. A middleware for intelligent environments and the internet of things. In *ACM UIC'10*, 2010.

[72] EPCglobal, www.epcglobalinc.org/.

[73] OPC Unified Architecture (UA), www.opcfoundation.org/ua/.

[74] ETSI machine-to-machine communications (m2m); functional architecture, draft ETSI TS 102 690 v0.10.4 (2011-01).

[75] 3GPP TS 22.368 v11.0.0, service requirements for machine-type communications.

[76] UPnP, http://www.upnp.org.

[77] Roy Thomas Fielding. Architectural styles and the design of network-based software architectures. Doctoral dissertation, University of California, Irvine, USA, 2000.

[78] Internet-of-things architecture (IOT-A) project deliverable d3.1 - initial m2m api analysis.

[79] Angelo Paolo Castellani, Nicola Bui, Paolo Casari, Michele Rossi, Zach Shelby, and Michele Zorzi. Architecture and protocols for the internet of things: A case study. In *IEEE PerCom Workshops'10*, pages 678–683, 2010.

[80] IPv6 over Low power WPAN (6lowpan), IETF. http://datatracker.ietf.org/wg/6lowpan/charter.

[81] Bilhanan Silverajan and Jarmo Harju. Developing network software and communications protocols towards the internet of things. In *IEEE COMSWARE '09*, pages 9:1–9:8, 2009.

[82] Dominique Guinard, Vlad Trifa, and Erik Wilde. A resource oriented architecture for the web of things. In *IEEE IoT'10*, 2010.

[83] N. Kong, N. Crespi, G. Lee, and J. Park. Internet-draft: The internet of things - concept and problem statement, 18 Oct. 2010. https://tools.ietf.org/html/draft-lee-iot-problem-statement-00.

[84] M. Y.S. Uddin and M. M. Akbar. Addressing techniques in wireless sensor networks: A short survey. In *IEEE ICECE'06*, pages 581–584, Dec. 2006.

[85] M. Ali and Z.A. Uzmi. An energy-efficient node address naming scheme for wireless sensor networks. In *IEEE INCC'04*, pages 25–30, June 2004.

[86] Fujun Ye and Ruifang Pan. A survey of addressing algorithms for wireless sensor networks. In *IEEE WiCom'09*, pages 1–7, Sept. 2009.

[87] Zhigao Du, Cui Zhang, Yangming Su, Depei Qian, and Yi Liu. Two-tier dynamic address assignment in wireless sensor networks. In *IEEE TENCON'09*, pages 1–6, Jan. 2009.

[88] Curt Schurgers, Gautam Kulkarni, and Mani B. Srivastava. Distributed assignment of encoded MAC addresses in sensor networks. In *ACM MobiHoc'01*, pages 295–298, 2001.

[89] Rodrigo Fonseca, Sylvia Ratnasamy, Jerry Zhao, Cheng Tien Ee, David Culler, Scott Shenker, and Ion Stoica. Beacon vector routing: Scalable point-to-point routing in wireless sensornets. In *USENIX NSDI'05*, 2005.

[90] F. D. Kronewitter. Dynamic Huffman addressing in wireless sensor networks based on the energy map. In *IEEE MILCOM'08*, pages 1–6, Nov. 2008.

[91] J. Elson and D. Estrin. Random, ephemeral transaction identifiers in dynamic sensor networks. In *IEEEE ICDCS'01*, pages 459–468, Apr. 2001.

[92] John Heidemann, Fabio Silva, Chalermek Intanagonwiwat, Ramesh Govindan, Deborah Estrin, and Deepak Ganesan. Building efficient wireless sensor networks with low-level naming. In *ACM Symp. on Oper. Sys. Principles (SOSP'01)*, pages 146–159, 2001.

[93] Qing Shen, Yu Liu, Zhijun Zhao, Song Ci, and Hui Tang. Distributed hash table based ID management optimization for internet of things. In *IEEE IWCMC'10*, pages 686–690, 2010.

[94] IBM Websphere MQ, www.ibm.com/software/integration/wmq/.

[95] Apache JMeter, http://jmeter.apache.org/.

[96] P. Ferguson and D. Senie. Network Ingress Filtering: Defeating Denial of Service Attacks Which Employ IP Source Address Spoofing.

[97] D.K.Y. Yau, J.C.S. Lui, Feng Liang, and Yeung Yam. Defending against distributed denial-of-service attacks with max-min fair server-centric router throttles. *IEEE/ACM Trans. on Networking*, 13(1):29–42, Feb. 2005.

[98] J. Ioannidis and S. M. Bellovin. Implementing pushback: Router-based defense against DDoS attacks. In *NDSS'02*.

[99] Zhiqiang Gao and N. Ansari. Tracing cyber attacks from the practical perspective. *IEEE Comm. Mag.*, 43(5):123–131, May 2005.

[100] Haining Wang, Danlu Zhang, and Kang G. Shin. Detecting SYN flooding attacks. In *IEEE INFOCOM'02*, volume 3, pages 1530–1539, June 2002.

[101] du Telecom, http://www.du.ae.

[102] Jennifer L. Wong and M. Potkonjak. Search in sensor networks: Challenges, techniques, and applications. In *Acoustics, Speech, and Signal Processing (ICASSP), 2002 IEEE International Conference on*, volume 4, pages IV–3752–IV–3755, 2002.

[103] Valerie Issarny, Mauro Caporuscio, and Nikolaos Georgantas. A perspective on the future of middleware-based software engineering. In *2007 Future of Software Engineering*, FOSE '07, pages 244–258, Washington, DC, USA, 2007. IEEE Computer Society.

[104] Casaleggio Associati. The evolution of internet of things. Technical report, Casaleggio Associati, February 2011. `http://www.casaleggio.it/pubblicazioni/Focus_internet_of_things_v1.81%20-%20eng.pdf` [Accessed on: 2011-06-08].

[105] European Commission. Internet of things in 2020 road map for the future. Technical report, Working Group RFID of the ETP EPOSS, May 2008. `http://ec.europa.eu/information_society/policy/rfid/documents/iotprague2009.pdf` [Accessed on: 2011-06-12].

[106] Patrick Guillemin and Peter Friess. Internet of things strategic research roadmap. Technical report, The Cluster of European Research Projects, September 2009. http://www.internet-of-things-research.eu/pdf/IoT_Cluster_Strategic_Research_Agenda_2009.pdf.

[107] International Telecommunication Union. Itu internet reports 2005: The internet of things. Workshop report, International Telecommunication Union, November 2005. http://www.itu.int/dms_pub/itu-s/opb/pol/S-POL-IR.IT-2005-SUM-PDF-E.pdf [Accessed on: 2011-12-12].

[108] Michal Nagy, Artem Katasonov, Oleksiy Khriyenko, Sergiy Nikitin, Michal Szydlowski, and Vagan Terziyan. Challenges of middleware for the internet of things. Technical report, University of Jyvaskyla, 2009. http://cdn.intechopen.com/pdfs/8786/InTech-Challenges_of_middleware_for_the_internet_of_things.pdf [Accessed on: 2011-12-20].

[109] Daqiang Zhang, L.T. Yang, and Hongyu Huang. Searching in internet of things: Vision and challenges. In *Parallel and Distributed Processing with Applications (ISPA), 2011 IEEE 9th International Symposium on*, pages 201–206, 2011.

[110] Xin Jin, Daqiang Zhang, Qin Zou, Genlin Ji, and Xiaojun Qian. Where searching will go in internet of things? In *Wireless Days (WD), 2011 IFIP*, pages 1–3, 2011.

[111] Simon Jirka, Arne Broring, and Christoph Stasch. Discovery mechanisms for the sensor web. *Sensors*, 9(4):2661–2681, 2009.

[112] Simon Mayer, Dominique Guinard, and Vlad Trifa. Searching in a web-based infrastructure for smart things. In *Proceedings of the 3rd International Conference on the Internet of Things (IoT 2012)*, Wuxi, China, 2012.

[113] S. Shakkottai. Asymptotics of query strategies over a sensor network. In *INFOCOM 2004. Twenty-third AnnualJoint Conference of the IEEE Computer and Communications Societies*, volume 1, 2004.

[114] Cuong Truong, Kay Romer, and Kai Chen. Fuzzy-based sensor search in the web of things. In *Proceedings of the 3rd International Conference on the Internet of Things (IoT 2012)*, Wuxi, China, 2012.

[115] B. Maryam Elahi, Kay Romer, Benedikt Ostermaier, Michael Fahrmair, and Wolfgang Kellerer. Sensor ranking: A primitive for efficient content-based sensor search. In *Proceedings of the 2009 International Conference on Information Processing in Sensor Networks*, IPSN '09, pages 217–228, Washington, DC, USA, 2009. IEEE Computer Society.

[116] B. Ostermaier, K. Römer, L.F. Mattern, M. Fahrmair, and W. Kellerer. A real-time search engine for the web of things. In *Proceedings of the 2rd International Conference on the Internet of Things (IoT 2010)*, pages 1–8, 2010.

[117] K. Romer, B. Ostermaier, F. Mattern, M. Fahrmair, and W. Kellerer. Real-time search for real-world entities: A survey. *Proceedings of the IEEE*, 98(11):1887–1902, 2010.

[118] H. Noguchi, T. Mori, and Tomomasa Sato. Framework for search application based on time segment of sensor data in home environment. In *Networked Sensing Systems (INSS), 2010 Seventh International Conference on*, pages 261–264, 2010.

[119] Haodong Wang, C.C. Tan, and Qun Li. Snoogle: A search engine for pervasive environments. *Parallel and Distributed Systems, IEEE Transactions on*, 21(8):1188–1202, 2010.

[120] Arne Broring, Felix Bache, Thomas Bartoschek, and Corne P.J.M. Elzakker. The sid creator: A visual approach for integrating sensors with the sensor web. In Stan Geertman, Wolfgang Reinhardt, and Fred Toppen, editors, *Advancing Geoinformation Science for a Changing World*, Lecture Notes in Geoinformation and Cartography, pages 143–162. Springer Berlin Heidelberg, 2011.

[121] OpenIoT Consortium. Open source solution for the internet of things into the cloud, January 2012. `http://www.openiot.eu` [Accessed on: 2012-04-08].

[122] w3.org. Semantic sensor network xg final report: W3c incubator group report, June 2011. `http://www.w3.org/2005/Incubator/ssn/XGR-ssn-20110628/` [Accessed on: 2012-09-25].

[123] Ioannis Paparrizos, Hoyoung Jeung, and Karl Aberer. Advanced search, visualization and tagging of sensor metadata. In *Proceedings of the 2011 IEEE 27th International Conference on Data Engineering*, ICDE '11, pages 1356–1359, Washington, DC, USA, 2011. IEEE Computer Society.

[124] Chiu C. Tan, Bo Sheng, Haodong Wang, and Qun Li. Microsearch: A search engine for embedded devices used in pervasive computing. *ACM Trans. Embed. Comput. Syst.*, 9(4):43:1–43:29, April 2010.

[125] Charith Perera, Arkady Zaslavsky, Peter Christen, and Dimitrios Georgakopoulos. Context aware computing for the internet of things: A survey. *Communications Surveys Tutorials, IEEE*, XX:X, 2013.

[126] Payam Barnaghi, Wei Wang, Cory Henson, and Kerry Taylor. Semantics for the internet of things: Early progress and back to the future. *Int. J. Semant. Web Inf. Syst.*, 8(1):1–21, January 2012.

[127] Digital Enterprise Research Institute. Linked sensor middleware (lsm), 2011. http://lsm.deri.ie/ [Accessed on: 2012-09-24].

[128] Danh Le Phuoc, Hoan Nguyen Mau Quoc, Josiane Xavier Parreira, and Manfred Hauswirth. The linked sensor middleware — connecting the real world and the semantic web. In *International Semantic Web Conference (ISWC)*, October 2011.

[129] Suman Nath, Jie Liu, and Feng Zhao. Sensormap for wide-area sensor webs. *Computer*, 40(7):90–93, July 2007.

[130] Charith Perera, Arkady Zaslavsky, Peter Christen, Michael Compton, and Dimitrios Georgakopoulos. Context-aware sensor search, selection and ranking model for internet of things middleware. In *IEEE 14th International Conference on Mobile Data Management (MDM)*, Milan, Italy, June 2013.

[131] John Domingue and Dieter Fensel. Toward a service web: integrating the semantic web and service orientation. *IEEE Intelligent Systems*, 23(1):8688, 2009.

[132] Suparna De, Tarek Elsaleh, Payam Barnaghi, and Stefan Meissner. An internet of things platform for real-world and digital objects. *Scalable Computing: Practice and Experience*, 13(1):45–57, 2012.

[133] D. Guinard, V. Trifa, S. Karnouskos, P. Spiess, and D. Savio. Interacting with the SOA-based internet of things: Discovery, query, selection, and on-demand provisioning of web services. *Services Computing, IEEE Transactions on*, 3(3):223–235, 2010.

[134] Talal Ashraf Butt, Iain Phillips, Lin Guan, and George Oikonomou. Trendy: An adaptive and context-aware service discovery protocol for 6LoWPANs. In *Proceedings of the Third International Workshop on the Web of Things*, WOT '12, pages 2:1–2:6, New York, NY, USA, 2012. ACM.

[135] Z. Shelby. Embedded web services. *Wireless Communications, IEEE*, 17(6):52–57, 2010.

[136] Jean-Paul Calbimonte, Hoyoung Jeung, Oscar Corcho, and Karl Aberer. Enabling query technologies for the semantic sensor web. *Int. J. Semant. Web Inf. Syst.*, 8(1):43–63, January 2012.

[137] Rau Garcia-Castro, Oscar Corcho, and Chris Hill. A core ontological model for semantic sensor web infrastructures. *Int. J. Semant. Web Inf. Syst.*, 8(1):22–42, January 2012.

[138] Naoya Namatame, Yong Ding, Till Riedel, Hideyuki Tokuda, Takashi Miyaki, and Michael Beigl. A distributed resource management architecture for interconnecting web-of-things using ubox. In *Proceedings of the Second International Workshop on Web of Things*, WoT '11, pages 4:1–4:6, New York, NY, USA, 2011. ACM.

[139] Joon Ahn and B. Krishnamachari. Modeling search costs in wireless sensor networks. In *Modeling and Optimization in Mobile, Ad Hoc and Wireless Networks and Workshops, 2007. WiOpt 2007. 5th International Symposium on*, pages 1–6, 2007.

[140] Zhiming Ding, Xu Gao, Limin Guo, and Qi Yang. A hybrid search engine framework for the internet of things based on spatial-temporal, value-based, and keyword-based conditions. In *Green Computing and Communications (GreenCom), 2012 IEEE International Conference on*, pages 17–25, 2012.

[141] Charith Perera, Arkady Zaslavsky, Peter Christen, and Dimitrios Georgakopoulos. Ca4iot: Context awareness for internet of things. In *IEEE International Conference on Conference on Internet of Things (iThing)*, pages 775–782, Besanon, France, November 2012.

[142] Commonwealth Scientific and Industrial Research Organisation (CSIRO), Australia. Phenonet: Distributed sensor network for phenomics supported by high resolution plant phenomics centre, CSIRO ICT centre, and CSIRO sensor and sensor networks tcp., 2011. `http://phenonet.com` [Accessed on: 2012-04-20].

[143] Christian Bizer and Andreas Schultz. The Berlin SPARQL benchmark. *Int. J. Semantic Web Inf. Syst.*, 5(2):1–24, 2009.

[144] Shuping Ran. A model for web services discovery with QoS. *SIGecom Exch.*, 4(1):1–10, March 2003.

[145] D. Chalmers and M. Sloman. A survey of quality of service in mobile computing environments. *Communications Surveys Tutorials, IEEE*, 2(2):2 –10, 1999.

[146] Gregory D. Abowd, Anind K. Dey, Peter J. Brown, Nigel Davies, Mark Smith, and Pete Steggles. Towards a better understanding of context and context-awareness. In *Proceedings of the 1st International Symposium on Handheld and Ubiquitous Computing*, HUC '99, pages 304–307, London, UK, 1999. Springer-Verlag.

[147] Apache Foundation. Commons math: The apache commons mathematics library, 2011. `http://commons.apache.org/math/` [2012-09-05].

[148] Apache Software Foundation. Apache Jena, November 2010. `http://jena.apache.org/` [accessed on: 2012-05-10].

[149] Australian Government, Bureau of Meteorology. Experimental environmental linked-data published by the Bureau of Meteorology, 2012. http://lab.environment.data.gov.au/ [Accessed on: 2012-009-05].

[150] Rick Cattell. Scalable SQL and NOSQL data stores. *SIGMOD Rec.*, 39(4):12–27, May 2011.

[151] Prasad Kulkarni. Distributed SPARQL query engine using MapReduce. Master's thesis, University of Edinburgh, 2010.

[152] Charith Perera, Arkady Zaslavsky, Michael Compton, Peter Christen, and Dimitrios Georgakopoulos. Semantic-driven configuration of internet of things middleware. In *9th International Conference on Semantics, Knowledge & Grids (SKG)*, page n/a, Beijing, China, October 2013.

[153] I.F. Akyildiz, Weilian Su, Y. Sankarasubramaniam, and E. Cayirci. A survey on sensor networks. *IEEE Comm. Mag.*, 40(8):102–114, Aug. 2002.

[154] C. Bisdikian, L. M. Kaplan, M. B. Srivastava, D. J. Thornley, D. Verma, and R. I. Young. Building principles for a quality of information specification for sensor information. In *FUSION 2009*, July.

[155] Richard Y. Wang and Diane M. Strong. Beyond accuracy: what data quality means to data consumers. *J. Manage. Inf. Syst.*, 12(4):5–33, 1996.

[156] M.E. Johnson and K.C. Chang. Quality of information for data fusion in net centric publish and subscribe architectures. In *FUSION'05*, July 2005.

[157] Z. Sun, C. H. Liu, C. Bisdikian, J. W. Branch, and B. Yang. QoI-aware energy management in Internet-of-Things sensory environments. In *IEEE SECON'12*, pages 19–27, June 2012.

[158] C. Bisdikian, J. Branch, K. K. Leung, and R. I. Young. A letter soup for the quality of information in sensor networks. In *IEEE PERCOM'09*, Galveston, TX, USA, March 2009.

[159] C. H. Liu, P. Hui, J. W. Branch, and Bo Yang. QoI-aware energy management for wireless sensor networks. In *IEEE PERCOM Workshops 2011 on IQ2S*, pages 8–13, March 2011.

[160] C. H. Liu, C. Bisdikian, J. W. Branch, and K. K. Leung. QoI-Aware wireless sensor network management for dynamic multi-task operations. In *IEEE SECON'10*, Boston, MA, USA, June 2010.

[161] J. Ma, W. Lou, Y. Wu, X. Li, and G. Chen. Energy efficient TDMA sleep scheduling in wireless sensor networks. In *IEEE INFOCOM'09*, pages 630–638, April 2009.

[162] Y. Wu, X. Li, Y. Liu, and W. Lou. Energy-efficient wake-up scheduling for data collection and aggregation. *IEEE Trans. Parallel and Distributed Systems*, 21(2):275–287, Feb. 2010.

[163] R. Jurdak, P. Baldi, and C. V. Lopes. Adaptive low power listening for wireless sensor networks. *IEEE Trans. Mobile Computing*, 6(8):988–1004, Aug. 2007.

[164] C. J. Merlin and W. B. Heinzelman. Duty cycle control for low-power-listening MAC protocols. *IEEE Trans. Mobile Computing*, 9(11):1508–1521, Nov. 2010.

[165] H. Yoo, M. Shim, and D. Kim. Dynamic duty-cycle scheduling schemes for energy-harvesting wireless sensor networks. *IEEE Comm. Letters*, 16(2):202–204, Feb. 2012.

[166] S. Zahedi, M. B. Srivastava, C. Bisdikian, and L. M. Kaplan. Quality tradeoffs in object tracking with duty-cycled sensor networks. In *IEEE RTSS'10*, pages 160–169, Nov. 2010.

[167] W. Ye, J. Heidemann, and D. Estrin. An energy-efficient MAC protocol for wireless sensor networks. 4(44):115–121, April 2006.

[168] T. Dam and K. Langendoen. An adaptive energy-efficient MAC protocol for wireless sensor networks. In *ACM SenSys'03*, pages 171–180, Oct. 2003.

[169] G. Tychogiorgos and C. Bisdikian. Seeking provides of relevant sensory information. In *IEEE MDM'11*, June 2011.

[170] E. Nakamura, F. Nakamura, C. Figueiredo, and A. Loureiro. Using information fusion to assist data dissemination in wireless sensor networks. *Telecom. Systems*, 30:237–254, 2005.

[171] E. Nakamura, A. Loureiro, and A. C. Frery. Information fusion for wireless sensor networks: Methods, models, and classifications. *ACM Comput. Surv.*, 39(3), September 2007.

[172] R. Pyke. Markov renewal processes: definitions and preliminary properties. *Ann. Math. Statist*, 32:1231–1242, 1961.

[173] A. P. Dempster, N. M. Laird, and D. B. Rubin. Maximum likelihood from incomplete data via the EM algorithm. *Journal of the Royal Statistical Society, Series B*, 39(1):1–38, 1977.

[174] ZigBee Alliance. ZigBee home automation public application profile. *IEEE J. Select. Areas Commun.*, Oct. 2007.

[175] Z-Wave. Z-wave protocol overview. May 2007.

[176] P. Darbee. Whitepaper: The details. Insteon, Aug. 2005.

[177] A. Garcia-Hernando et al. *Problem solving for wireless sensor networks.* Springer, July 2008.

[178] Routing Over Low power and Lossy networks (roll), IETF. *Available at: http://datatracker.ietf.org/wg/roll/charter.*

[179] Constrained RESTful Environments (core), IETF. *Available at: http://datatracker.ietf.org/wg/core/charter.*

[180] IP Smart Object Alliance (IPSO). *Available at: http://www.ipso-alliance.org.*

[181] Cesare Pautasso, Olaf Zimmermann, and Frank Leymann. Restful web services vs. "big" web services: Making the right architectural decision. In *17th International Conference on World Wide Web*, 2008.

[182] Weijun Qin, Qiang Li, Limin Sun, Hongsong Zhu, and Yan Liu. Restthing: A restful web service infrastructure for mash-up physical and web resources. In *2011 IFIP 9th International Conference on Embedded and Ubiquitous Computing (EUC)*, pages 197–204, Oct. 2011.

[183] V. Stirbu. Towards a restful plug and play experience in the web of things. In *IEEE International Conference on Semantic Computing*, pages 512–517, Aug. 2008.

[184] Dominique Guinard. Towards the web of things: Web mashups for embedded devices. In *In MEM 2009 in Proceedings of WWW 2009. ACM*, 2009.

[185] D. Guinard, V. Trifa, and E. Wilde. A resource oriented architecture for the web of things. In *Internet of Things (IOT), 2010*, Dec. 2010.

[186] T. Potsch, K. Kuladinithi, M. Becker, P. Trenkamp, and C. Goerg. Performance evaluation of CoAP using RPL and LPL in TinyOs. In *2012 5th International Conference on New Technologies, Mobility and Security (NTMS)*, pages 1–5, May 2012.

[187] S. Sitharama Iyengar, Nandan Parameshwaran, Vir V. Phoha, N. Balakrishnan, and Chuka D. Okoye. *Fundamentals of sensor network programming: Applications and technology.* Wiley-IEEE Press, 2011.

[188] A. Dunkels, B. Gronvall, and T. Voigt. Contiki — a lightweight and flexible operating system for tiny networked sensors. In *29th Annual IEEE International Conference on Local Computer Networks, 2004*, pages 455– 462, Nov. 2004.

[189] M. Kovatsch, S. Duquennoy, and A. Dunkels. A low-power CoAP for Contiki. In *2011 IEEE 8th International Conference on Mobile Adhoc and Sensor Systems (MASS)*, pages 855–860, Oct. 2011.

[190] Jürgen Schönwalder, Tina Tsou, and Behcet Sarikaya. Protocol profiles for constrained devices. *Available at: www.iab.org/wp-content/IAB-uploads/.../Schoenwaelder.pdf*, 2011.

[191] IEEE Computer Society. IEEE std. 802.15.4-2003. 2003.

[192] G. Montenegro, N. Kushalnagar, J. Hui, and D. Culler. Transmission of IPv6 packets over ieee 802.15.4 networks. *RFC4944, available at: https://datatracker.ietf.org/doc/rfc4944/.*

[193] J. Hui and P. Thubert. Compression format for ipv6 datagrams over ieee 802.15.4-based networks. *RFC6282, available at: hhttp://datatracker.ietf.org/doc/rfc6282/.*

[194] P. Levisi, A. Tavakoli, and S. Dawson-Haggerty. Overview of existing routing protocols for low power and lossy networks. *Internet-Draft, available at: http://tools.ietf.org/html/draft-ietf-roll-protocols-survey-07.*

[195] J.P. Vasseuri, M. Kim, K. Pister, N. Dejean, and D. Barthel. Routing metrics used for path calculation in low-power and lossy networks. *RFC 6551, available at: http://datatracker.ietf.org/doc/rfc6551/.*

[196] T. Winter, P. Thubert, A. Brandt, J. Hui, R. Kelsey, P. Levis, K. Pister, R. Struik, JP. Vasseur, and R. Alexander. Rpl: Ipv6 routing protocol for low-power and lossy networks. *RFC 6550, available at: http://datatracker.ietf.org/doc/rfc6550/.*

[197] O. Bergmann. libcoap: C-implementation of CoAP. *Available at: http://libcoap.sourceforge.net.*

[198] Copper (Cu) CoAP Browser. A Firefox add-on to browse the internet of things. *Available at: https://github.com/mkovatsc/Copper.*

[199] H. Jin, L. Ding, J. Yu, D. Niu, F. Liu, P. Shu, and B. Li. Gearing resource-poor mobile devices with powerful clouds: architectures, challenges, and applications. *IEEE Wirel. Comm.*, 20:2–10, June 2013.

[200] S. W. Loke, N. Fernando, and W. Rahayu. Mobile cloud computing: A survey. *Fut. Gen. Comp. Sys.*, 29:84–106, June 2013.

[201] M. Satyanarayanan. Mobile computing: the next decade. *SIGMOBILE Mob. Comput. Commun. Rev.*, 15:2–10, Aug. 2011.

[202] J. Dean and S. Ghemawat. MapReduce: simplified data processing on large clusters. *ACM Commun.*, 51:107–113, Jan. 2008.

[203] E. Marinelli. *Hyrax: Cloud computing on mobile devices using MapReduce.* PhD thesis, Carnegie Mellon University, 2009.

[204] R. Caceres, M. Satyanarayanan, P. Bahl, and N. Davies. The case for VM-based cloudlets in mobile computing. *IEEE Pervasive Computing*, 8(4):14–23, 2009.

[205] D. Cho, A. Wolman, S. Saroiu, R. Chandra, E. Cuervo, A. Balasubramanian, and P. Bahl. Maui: Making smartphones last longer with code offload. In *ACM MobiSys'10*, pages 49–62, 2010.

[206] M. Kang, D. Huang, X. Zhang, and J. Luo. Mobicloud: Building secure cloud framework for mobile computing and communication. In *IEEE SOSE'10*, pages 27–34, 2010.

[207] M. D. Kristensen. Scavenger: Transparent development of efficient cyber foraging applications. In *IEEE PerCom'10*, pages 217–226, 2010.

[208] P. Maniatis, M. Naik, B.-G. Chun, S. Ihm, and A. Patti. Clonecloud: elastic execution between mobile device and cloud. In *ACM EuroSys '11*, pages 301–314, 2011.

[209] P. Hui, R. Mortier, S. Kosta, A. Aucinas, and X. Zhang. Thinkair: Dynamic resource allocation and parallel execution in the cloud for mobile code offloading. In *IEEE INFOCOM'12*, pages 945–953, 2012.

[210] S. Mehrotra M. R. Rahimi, N. Venkatasubramanian and A. V. Vasilakos. Mapcloud: Mobile applications on an elastic and scalable 2-tier cloud architecture. In *IEEE/ACM UCC'12*, 2012.

[211] T. Kielmann, R. Kemp, N. Palmer, and H. Bal. Cuckoo: A computation offloading framework for smartphones. *Mob. Comp., App., and Serv.*, 76:59–79, 2012.

[212] S.-J. Lee, K.-H. Kim, and P. Congdon. On cloud-centric network architecture for multi-dimensional mobility. *ACM SIGCOMM Comp. Comm. Rev.*, 42:509–514, 2012.

[213] Y.-H. Lu, K. Kumar, J. Liu, and B. Bhargava. A survey of computation offloading for mobile systems. *ACM/Springer MONET*, 18:129–140, Apr. 2013.

[214] K. Kumar and Y.-H. Lu. Cloud computing for mobile users: Can offloading computation save energy? *IEEE Computer*, 43(4):51–56, Apr. 2010.

[215] Y.-H. Lu, Y. Nimmagadda, K. Kumar, and C. S. G. Lee. Real-time moving object recognition and tracking using computation offloading. In *IEEE/RSJ Intelligent Robots and Systems (IROS'10)*, pages 2449–2455, 2010.

[216] S. Ou, K. Yang, and H.-H. Chen. On effective offloading services for resource-constrained mobile devices running heavier mobile internet applications. *IEEE Comm. Mag.*, 46:56–63, Jan. 2008.

[217] C. Krintz, R. Wolski, S. Gurun, and D. Nurmi. Using bandwidth data to make computation offloading decisions. In *IEEE International Symposium on Parallel and Distributed Processing*, Apr. 2008.

[218] K. Kumar, Y.-J. Hong, and Y.-H. Lu. Energy efficient content-based image retrieval for mobile systems. In *IEEE ISCAS'09*, pages 1673–1676, 2009.

[219] R. Nathuji, B. Seshasayee, and K. Schwan. Energy-aware mobile service overlays: Cooperative dynamic power management in distributed mobile systems. In *IEEE ICAC '07*, 2007.

[220] G. Huerta-Canepa and D. Lee. An adaptable application offloading scheme based on application behavior. In *IEEE AINAW'08 Wkshp*, pages 387–392, 2008.

[221] Y.-H. Lu, C. Xian, and Z. Li. Adaptive computation offloading for energy conservation on battery-powered systems. In *Parallel & Dist. Sys.'07*, 2007.

[222] K. Harras, A. Mtibaa, A. Fahim, and M. Ammar. Towards resource sharing in mobile device clouds: Power balancing across mobile devices. In *ACM SIGCOMM MCC'13*, pages 51–56, 2013.

[223] A. Gani, Z. Sanaei, S. Abolfazli, and R. Buyya. Heterogeneity in mobile cloud computing: Taxonomy and open challenges. *IEEE Comm. Surv. & Tut.*, (99):1–24, 2013.

[224] X. Liu, J. Li, K. Bu, and B. Xiao. Enda: embracing network inconsistency for dynamic application offloading in mobile cloud computing. In *ACM SIGCOMM Workshop on MCC'13*, pages 39–44, 2013.

[225] J. Schneider, A. Klein, C. Mannweiler, and H. D. Schotten. Access schemes for mobile cloud computing. In *IEEE MDM'10*, pages 387–392, 2010.

[226] S. Subashini and V. Kavitha. A survey on security issues in service delivery models of cloud computing. *J. Netw. Comput. App.*, 34(1):1–11, 2011.

[227] Q. Li and G. Clark. A survey on security issues in service delivery models of cloud computing. *J. Netw. Comput. App.*, 11(1):78–81, 2013.

[228] S. U. Khan, A. N. Khan, M. L. M. Kiah, and S. A. Madani. Towards secure mobile cloud computing: A survey. *Fut. Gen. Comp. Sys.*, 29(5):1278–1299, 2013.

[229] P. Kumar, S. S. Walia, J.-P. Hubaux, I. Aad I. Bilogrevic, M. Jadliwala, and V. Niemi. Meetings through the cloud: Privacy-preserving scheduling on mobile devices. *J. Syst. & Softw.*, 84(11):1910–1927, 2011.

[230] A. Kayssi, W. Itani, and A. Chehab. Energy-efficient incremental integrity for securing storage in mobile cloud computing. In *IEEE ICEAC'10*, pages 21–26, 2010.

[231] L. X. Cai, X. Shen, H. Liang, D. Huang, and D. Peng. Resource allocation for security services in mobile cloud computing. In *IEEE INFOCOM'11 WKSHPS on M2MCN'11*, pages 191–195, 2011.

[232] X. Huang, M. L. Yiu, C. S. Jensen, and H. Lu. Spacetwist: Managing the trade-offs among location privacy, query performance, and query accuracy in mobile services. In *IEEE ICDE'08*, pages 366–375, 2008.

[233] D. Huang, Y. Zhu, D. Ma, and C. Hu. Enabling secure location-based services in mobile cloud computing. In *ACM SIGCOMM Workshop on MCC'13*, pages 27–32, 2013.

[234] E. Hossain, W. Saad, D. Niyato, P. Wang, and Z. Han. Game theoretic modeling of cooperation among service providers in mobile cloud computing environments. In *IEEE WCNC'12*, pages 3128–3133, 2012.

[235] D. Huang, X. Shen, H. Liang, L. X. Cai, and D. Peng. An smdp-based service model for interdomain resource allocation in mobile cloud networks. *IEEE Transactions on Vehicular Technology*, 61(5):2222–2232, 2012.

[236] M. P. Papazoglou. Cloud blueprints for integrating and managing cloud federations. *Software Service and Application Engineering, LNCS*, 7365(5):102–119, 2012.

[237] S. Maiti, S. Bandyopadhyay, M. Sengupta, and S. Dutta. A survey of middleware for internet of things. *Recent Trends in Wireless and Mobile Networks*, pages 288–296, 2011.

[238] A. M. Chaqfeh and N. Mohamed. Challenges in middleware solutions for the internet of things. In *IEEE International Conference on Collaboration Technologies and Systems*, pages 21–26, 2012.

[239] I. Carreras, D. Miorandi, A. Tamilin, E.R. Ssebaggala, and N. Conci. Crowd-sensing: Why context matters. In *Pervasive Computing and Communications Workshops (PERCOM Workshops), 2013 IEEE International Conference on*, pages 368–371, 2013.

[240] Hande Alemdar and Cem Ersoy. Wireless sensor networks for healthcare: A survey. *Computer Networks*, 54(15):2688–2710, 2010.

[241] V. Stanford. Using pervasive computing to deliver elder care. *IEEE Pervasive Comp.*, 1(1):10 – 13, 2002.

[242] T. Mcfadden and J. Indulska. Context-aware environments for independent living. In *IEEE 3rd National Conf. of Emerging Researchers in Ageing*, 2004.

[243] J. A. Stankovic, Q. Cao, T. Doan, L. Fang, Z. He, R. Kiran, S. Lin, S. Son, R. Stoleru, and A. Wood. Wireless sensor networks for in-home healthcare: Potential and challenges. In *IEEE High Confidence Medical Device Software and Systems (HCMDSS) Workshop'05*, 2005.

[244] V. Jones, R. Kleissen, V. V. Goldman, A. T. Halteren, I. A. Widya, and N. T. Dokovski. Mobile applications in the health sector. In *Mobile Minded Symp.*, volume 22, 1999.

[245] S. Sultan and P. Mohan. How to interact: Evaluating the interface between mobile healthcare systems and the monitoring of blood sugar and blood pressure. In *ACM MobiQuitous'09*, pages 1–6, July 2009.

[246] Yonglin Ren, R. W. N. Pazzi, and A. Boukerche. Monitoring patients via a secure and mobile healthcare system. *IEEE Trans. on Wireless Comm.*, 17(1):59–65, Feb. 2010.

[247] P. Mohan and S. Sultan. Medinet: A mobile healthcare management system for the caribbean region. In *ACM MobiQuitous'09*, pages 1–2, July 2009.

[248] G. Wu, S. Talwar, K. Johnsson, N. Himayat, and K. D. Johnson. M2M: From mobile to embedded internet. *IEEE Comm. Mag.*, 49(4):36–43, April 2011.

[249] M. Conner. Sensors empower the internet of things. *EDN*, 55:10–32, 2010.

[250] S. Haller, S. Karnouskos, and C. Schroth. The internet of things in an enterprise context. *Future Internet–FIS 2008*, pages 14–28, 2009.

[251] V. Jones, A. Halteren, I. Widya, N. Dokovsky, G. Koprinkov, R. Bults, D. Konstantas, and R. Herzog. Mobihealth: Mobile health services based on body area networks. *Springer M-Health*, pages 219–236, 2006.

[252] S. Shea, J. Starren, R. S. Weinstock, P. E. Knudson, J. Teresi, D. Holmes, W. Palmas, L. Field, R. Goland, and C. Tuck. Columbia University: Informatics for diabetes education and telemedicine (ideatel) project. *J. of the American Medical Informatics Association*, 9(1):49, 2002.

[253] T. Broens, A. Van-Halteren, M. Van-Sinderen, and K. Wac. Towards an application framework for context-aware m-health applications. *Int'l J. of Internet Protocol Tech.*, 2(2):109–116, 2007.

[254] Y. Zouand, R.S.H. Istepanian, and W. Huang. Performance evaluation of a GPRS/bluetooth diabetes management system. In *IET 3rd Int'l Conf. on Advances in Medical, Sig. and Inf. Processing'06*, pages 1–4, 2007.

[255] S. Dagtas, Y. Natchetoi, and H. Wu. An integrated wireless sensing and mobile processing architecture for assisted living and healthcare applications. In *ACM SIGMOBILE Int'l Workshop on Sys. and Netw. Support for Healthcare and Assisted Living Environments*, pages 70–72, 2007.

[256] R. Carroll, R. Cnossen, M. Schnell, and D. Simons. Continua: An interoperable personal healthcare ecosystem. *IEEE Pervasive Comp.*, pages 90–94, 2007.

[257] L. Richardson and S. Ruby. *Restful web services.* O'Reilly Media, May 2007.

[258] Roy T. Fielding and Richard N. Taylor. Principled design of the modern web architecture. In *ACM ICSE'00*, pages 407–416, 2000.

[259] D. Guinard and V. Trifa. Towards the web of things: Web mashups for embedded devices. In *WWW Workshop on Mashups, Enterprise Mashups and Lightweight Composition on the Web (MEM'09)*, 2009.

[260] MQTT. http://mqtt.org.

[261] Glenn Lyons and Kiron Chatterjee. A human perspective on the daily commute: Costs, benefits and trade-offs. *Transport Reviews*, 28(2):181–198, 2008.

[262] Youngki Lee, Younghyun Ju, Chulhong Min, Seungwoo Kang, Inseok Hwang, and Junehwa Song. CoMon: Cooperative ambience monitoring platform with continuity and benefit awareness. In *ACM MobiSys'12*, pages 43–56, 2012.

[263] Carrie Heeter, Chandan Sarkar, Becky Palmer-Scott, and Shasha Zhang. Engineering sociability: Friendship drive, visibility, and social connection in anonymous co-located local Wi-Fi multiplayer online gaming. *Int'l J. of Gaming and Computer-Mediated Simulations*, 4(2):1–18, 2012.

[264] Jason O. B. Soh and Bernard C. Y. Tan. Mobile gaming. *Commun. ACM*, 51(3):35–39, March 2008.

[265] Shaoxuan Wang and Sujit Dey. Cloud mobile gaming: modeling and measuring user experience in mobile wireless networks. *ACM SIGMOBILE Mob. Comput. Commun. Rev.*, 16(1):10–21, July 2012.

[266] Claudio Feijoo, José-Luis Gómez-Barroso, Juan-Miguel Aguado, and Sergio Ramos. Mobile gaming: Industry challenges and policy implications. *Telecommun. Policy*, 36(3):212–221, April 2012.

[267] P. E. Ross. Cloud computing's killer app: Gaming. *IEEE Spectrum*, 46(3):14, 2009.

[268] E. J. Berglund and D. R. Cheriton. Amaze: a multiplayer computer game. *IEEE Software*, 2(3):30–39, May 1985.

[269] Y. Bernier. Latency compensating methods in client/server in-game protocol design and optimization. In *Games Developers Conf.'01.*

[270] Andreea Picu and Thrasyvoulos Spyropoulos. Forecasting dtn performance under heterogeneous mobility: The case of limited replication. In *IEEE SECON'11*, pages 569–577, June 2011.

[271] Vikram Srinivasan, Mehul Motani, and Wei Tsang Ooi. Analysis and implications of student contact patterns derived from campus schedules. In *ACM MobiCom'06*, pages 86–97, 2006.

[272] Hong Lu, Wei Pan, Nicholas D. Lane, Tanzeem Choudhury, and Andrew T. Campbell. SoundSense: Scalable sound sensing for people-centric applications on mobile phones. In *ACM MobiSys'09*, pages 165–178, 2009.

[273] Lorenzo Keller, Anh Le, Blerim Cici, Hulya Seferoglu, Christina Fragouli, and Athina Markopoulou. MicroCast: cooperative video streaming on smartphones. In *ACM MobiSys'12*, pages 57–70, 2012.

[274] Hayoung Yoon and Jongwon Kim. Collaborative streaming-based media content sharing in wifi-enabled home networks. *IEEE Trans. on Consumer Electronics*, 56(4):2193–2200, 2010.

[275] M. G.-Domingo and J. A. M. Forner. Cube-u: Exploring the combination of the internet of things and elearning. In *The 2011 Annual Conference Extended Abstracts on Human Factors in Computing Systems*, 2011.

[276] M. Wang and J. W. P. Ng. Intelligent mobile cloud education: Smart anytime-anywhere learning for the next generation campus environment. In *IEEE Intelligent Environments (IE'12)*, pages 149–156, 2012.

[277] S. Chung, S. Wang, and I. Rhee. A cloud-powered driver-less printing system for smartphones. In *ACM UbiComp '13*, pages 255–264, 2013.

[278] T. K. Lakshman and X. Thuijs. Enhancing enterprise field productivity via cross platform mobile cloud apps. In *ACM MCS '11*, pages 27–32, 2011.

[279] M. Sihvonen M. R. Savola, H. Abie. Towards metrics-driven adaptive security management in e-health iot applications. In *The 7th International Conference on Body Area Networks*, pages 276–281, 2012.

[280] S. Ullah W. Cai M. Chen, Y. Ma and E. Song. Rochas: Robotics and cloud-assisted healthcare system for empty nester. In *BodyNets'13*, 2013.

[281] DARPA. Creating a secure, private internet and cloud at the tactical edge, 2013. http://www.darpa.mil/NewsEvents/Releases/2013/08/21.aspx [Online. Last accessed: 2013/09/20].

Index